Research and Evidence in Software Engineering

From Empirical Studies to Open Source Artifacts

Edited by
Varun Gupta
Chetna Gupta

CRC Press
Taylor & Francis Group
Boca Raton London New York

CRC Press is an imprint of the
Taylor & Francis Group, an **informa** business

AN AUERBACH BOOK

First Edition published 2021
by CRC Press
6000 Broken Sound Parkway NW, Suite 300, Boca Raton, FL 33487-2742
and by CRC Press
2 Park Square, Milton Park, Abingdon, Oxon, OX14 4RN

© 2021 Taylor & Francis Group, LLC
CRC Press is an imprint of Taylor & Francis Group, LLC

ISBN: 978-0-367-35852-5 (hbk)
ISBN: 978-0-367-76765-5 (pbk)
ISBN: 978-1-003-16839-3 (ebk)

Typeset in Adobe Garamond Pro
by KnowledgeWorks Global Ltd.

Contents

Preface vii
Editor Biographies xv
Contributor Biographies xvii

1 Performance of Execution Tracing with Aspect-Oriented and
   Conventional Approaches 1
   TAMAS GALLI, FRANCISCO CHICLANA, AND FRANCOIS SIEWE

2 A Survey on Software Test Specification Qualities for Legacy
   Software Systems 59
   SHILPA MARIAM GEORGE, SARADHA YASASVI YEDURUVADA,
   AND DAE-KYOO KIM

3 Whom Should I Talk To?: And How That Can Affect My Work 75
   SUBHAJIT DATTA

4 Software Project Management: Facts versus Beliefs and Practice 89
   LAWRENCE PETERS

5 Inter-Parameter Dependencies in Real-World Web APIs: The
   IDEA Dataset 101
   ALBERTO MARTIN-LOPEZ, SERGIO SEGURA,
   AND ANTONIO RUIZ-CORTÉS

6 Evaluating Testing Techniques in Highly-Configurable Systems:
   The Drupal Dataset 107
   ANA B. SÁNCHEZ, SERGIO SEGURA,
   AND ANTONIO RUIZ-CORTÉS

7 A Family of Experiments to Evaluate the Effects of Mindfulness
   on Software Engineering Students: The MetaMind Dataset 115
   BEATRIZ BERNÁRDEZ, MARGARITA CRUZ, AMADOR DURÁN,
   JOSÉ A. PAREJO, AND ANTONIO RUIZ-CORTÉS

8 Process Performance Indicators for IT Service Management: The PPI Dataset..125
BEDILIA ESTRADA-TORRES, ADELA DEL-RÍO-ORTEGA, MANUEL RESINAS, AND ANTONIO RUIZ-CORTÉS

9 Prioritization in Automotive Software Testing: Systematic Literature Review and Directions for Future Research.....................133
NAOHIKO TSUDA, ANKUSH DADWAL, HIRONORI WASHIZAKI, YOSHIAKI FUKAZAWA, MASASHI MIZOGUCHI, AND KENTARO YOSHIMURA

10 Deep Embedding of Open Source Software Bug Repositories for Severity Prediction 159
ABEER HAMDY AND GLORIA EZZAT

11 Predict Who: An Intelligent Game Using NLP and Knowledge Graph Model 179
TAMEEM AHMAD, NESAR AHMAD, MOHAMMAD SAQIB, AND ABU HUZAIFA KHAN

12 Mining Requirements and Design Documents in Software Repositories Using Natural Language Processing and Machine Learning Approaches 199
ISHAYA GAMBO, CLAVERS CHABI, SIMON YANGE, RHODA IKONO, AND THERESA OMODUNBI

13 Empirical Studies on Using Pair Programming as a Pedagogical Tool in Higher Education Courses: A Systematic Literature Review 251
KULJIT KAUR CHAHAL, AMANPREET KAUR, AND MUNISH SAINI

14 Programming Multi-Agent Coordination Using NorJADE Framework...287
TOUFIK MARIR, SELMA MAMMERI, AND ROHALLAH BENABOUD

Index 303

Preface

An Overview

Software Engineering deals with the delivery of high quality software to its users within time and budgets. New technologies, techniques, tools, and empirical evidence are evolving continuously as a result of ever-changing business environments, which bring higher levels of uncertainties in software development and release. The complexity of the software code is becoming more complex day by day and there is a pressure to release high quality code to diverse markets in the shortest time under a high level of uncertainties. The availability of the software engineering knowledge as reported through research studies, which are made accessible to software practitioners, helps to minimize the above-mentioned constraints leading to the growth of software engineering practices.

The software engineering research community is working on improving the software engineering practices, improving the quality of the developed software that provides the benefits to both the software company and the customers and/ or end users. The continuous innovations in software engineering are disseminated within the community in the form of rigorous empirical studies, validated solutions, evaluated solutions, open source tools, experience reports and opinions. This helps the community to learn from empirical findings and adopt the research solutions to their context specific scenarios. Disseminating the types of research contributions within the community helps to foster collaboration among the software engineering community. This collaboration is further expanded by sharing the primary research datasets and open source tools, which could be used as such by other researchers to expand their research or adopted as per their research needs.

Data driven software engineering is making it necessary to continuously manage the tremendous amount of large, heterogenous noisy data (for instance bug reports, customer feedbacks, historical data, etc.) to drive quality decision making in the companies. This requires the application of efficient techniques from other areas of computer science like Artificial Intelligence, Natural Language Processing, Probability & Statistics, Crowdsourcing, Big Data, Data Mining, etc.

This book provides relevant theoretical frameworks, empirical research findings and evaluated solutions addressing research challenges in the software engineering domain. The research contained in this book highlights issues, challenges, techniques and practices relevant to the software engineering research community. To foster collaboration between the software engineering research community, this book also reports the datasets related to various software engineering aspects as acquired systematically through scientific methods and are valuable to the research community. The datasets of primary research will allow other researchers to use them in their research, which improves the quality of the overall research. The knowledge disseminated by the research studies contained in the book will motivate other researchers to further innovate the way that software development happens in real practice.

Target Audience

The research findings contained in this book are ideals for software engineers working in industries, scientists working as researchers with research organizations, universities and research & development units, academicians, software consultants, management executives, etc. This book provides the audience with a single platform for getting access to recent research in the area of software engineering not only in terms of the research solutions, but also empirical evidences which help them to make decisions driven by stronger evidences. The research studies as reported in this book will help serve the purpose of its audience in following way:

Academicians

- Teaching latest software engineering studies to the students
- Using the material for innovating their research lines
- Formulating research proposals for submissions to the funding agencies

Scientists/Researchers

- Using the material for innovating their research lines
- Formulating research proposals for submissions to the funding agencies
- Innovating techniques, tools and practices as employed in their company context
- Using them as the basis for their research degrees as the primary literature studies

Software Engineers

- Adopting the research techniques, tools and practices in their working context to improve the way they use to work earlier
- Evaluating the studies in the industrial context and adopting them that best meet their needs

Management Executives

- Decision making related to the innovation of their existing strategies and policies about software development techniques, tools and practices employed
- Incorporating interdisciplinary knowledge and other companies' practices solving their unique business needs

Contributors Demographics

This book includes contributions from 46 authors from leading universities worldwide. These authors contributed to all 14 chapters of the book. The contributors came from different research organizations including Iowa State University (Iowa); Oakland University (Michigan); Universidad Politecnica de Madrid (Spain); University of Seville (Spain); University of Tartu (Estonia); De Montfort University (UK); Waseda University Tokyo (Japan); Research & Development Group, Hitachi Ltd. (Japan); Singapore Management University (Singapore); Guru Nanak Dev University (India); Aligarh Muslim University (India); British University in Egypt (Egypt); University of Oum El Bouaghi (Algeria); Obafemi Awolowo University (Nigeria); and Federal University of Agriculture (Nigeria) (Figure I.1).

CONTRIBUTOR DEMOGRAPHICS (CONTINENT WISE)

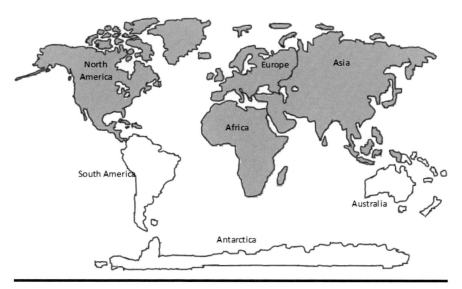

Figure I.1 Contributor demographics (on basis of their affiliations single/multiple).

Organisation of Book

This book is organized into 14 chapters and each chapter provides insight into software engineering-related aspects. The book presents interesting topics of the software engineering domain for its audience, as shown in Figure I.2.

The percentage distribution of the contributions into research types include:

- Out of fourteen studies, four studies (29% approximately) report the datasets related to various software engineering aspects as acquired systematically through scientific methods and are valuable to the research community.
- Six out of fourteen studies (43% approximately) are empirical research studies that report systematic literature reviews and experimentations.
- Four studies propose solutions to the software engineering research challenges (29% approximately). These studies are evaluated experimentally on live datasets. The major focus of these studies was to propose a new solution and, to prove its efficiency, it was evaluated experimentally.

These studies are housed in fourteen chapters which are individually described in the following paragraphs.

Chapter 1 reported an experimentation that aims to measure the performance of execution tracing with a resource intensive test application pair that implements

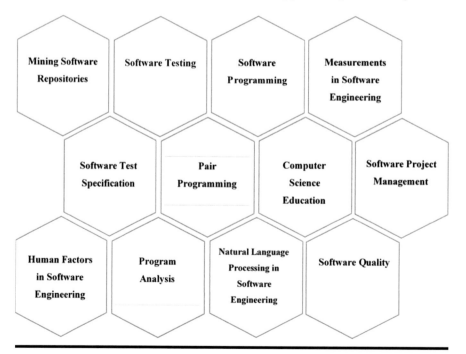

Figure I.2 Primary research topics addressed in the book.

the same functionality and produces the same amount of trace data. The one application possesses aspect-oriented execution tracing while the other one possesses conventional, object-oriented execution-tracing with manually inserted trace method calls. The performance impact of constructing the trace messages is measured by collecting the data in the applications. The performance impact of writing these data into a physical file is also measured.

Chapter 2 presents a survey study on test specification qualities. The survey studies the most concerned qualities for test specifications of legacy systems, other quality aspects that are considered as important in the field, the factors that affect the qualities of test specifications and field difficulties in developing quality test specifications. The survey was conducted among experienced industry experts from diverse domains with various technology backgrounds.

Chapter 3 examines whether and how the extent of collaboration between developers and the level of uniformity in their interaction relate to team outcomes. To achieve this objective, the data from a set of 389 real-world software projects which involve more than 200,000 developers, who collectively work on more than 300,000 defects, are examined statistically by testing formulated hypotheses.

Chapter 4 provides the readers with insights into the nature of software project management today based on facts and data as opposed to hearsay and marketing sales pitches. The method used to develop the work mentioned in this chapter was based on the literature survey of the relevant articles on the related topic.

Chapter 5 provides evidence on how inter-parameter dependencies are used in real-world web APIs. The documentation of 2,557 operations from 40 real-world web APIs was reviewed and carefully analyzed, and 633 inter-parameter dependencies were found and classified into seven different types.

Chapter 6 presents the Drupal dataset, a collection of real-world data collected from the popular open-source Drupal framework. This dataset allows the assessment of variability testing techniques with real data from a Highly-Configurable System.

Chapter 7 helps understand how various datasets related to the study of the impact of the mindfulness practice on students' performance in conceptual modeling tasks have been generated and trace their evolution across the family of experiments resulting in a single data set called MetaMind. Not only does the process include the integration of datasets into a single-target dataset with the complete set of data, but it also improves the structure of the dataset and adapts it to several statistical analysis tools.

Chapter 8 presents the process performance indicators dataset called PPI Dataset, which is defined in natural language and related to the IT service management processes of several public organizations in Spain. This dataset has been used as a basis for the development of techniques for modeling PPIs using different notations as well as for the transformation of natural language PPI definitions into others amenable to automated computation, applying natural language processing and matching learning techniques.

Chapter 9 reports the survey of the prioritization in automotive software testing to highlight a clear overview and devise a solid basis to plan for future research. A systematic literature review was conducted on the studies reported in IEEE Xplore, ACM Digital Library and Scopus bibliographic databases.

Chapter 10 reports an approach that utilizes word embedding and deep learning models for automatic identification of severity classes of newly submitted bug reports. The results are validated using two open large-scale bug repositories, i.e. Eclipse and Mozilla.

Chapter 11 proposes a web application "Predict Who" to program the game and let the system find out the personality the user is thinking of by asking him/her questions one by one. Natural Language Processing is used for retrieving user-specific information and the formulation of queries. Internally, the data is organized on the basis of a Knowledge Graph, which is an intelligent model of real-world entities that depicts their relations to one another as well.

Chapter 12 reports a recommendation system that will help software developers produce a new product of good quality by helping the software developers to locate useful data in a software development project without wasting time. The overarching goal is to understand and resolve the challenges, complexities and peculiarities of data in software repositories. The performance of the model is evaluated using recall, precision and execution time as performance parameters.

Chapter 13 reports a systematic literature review of the current empirical evidence of using pair programming as a pedagogical tool in Computer Science education in universities/colleges. A systematic literature review is conducted on the studies reported in five bibliographic databases including IEEE Explore, ACM digital library, Springer Link, Science Direct and Wiley online library. The selected studies are further subjected to the Snowballing approach to further extend the collection with more relevant studies.

Chapter 14 reports the use of the norms to ensure the coordination between agents in the multi-agent systems by extending the NorJADE framework. NorJADE framework is an extension of the JADE framework to develop normative multi-agent systems. The proposed extensions are made thanks to extending the ontology and implementing aspects to develop the coordination mechanisms.

This book contains research articles addressing numerous research challenges associated with different software development-related activities like programming, testing, measurements, human factors (social software engineering), specification, quality, program analysis, software project management, etc. The knowledge disseminated by this book will help the software engineering community working in different software engineering areas to benefit from the reported research in terms of its ability to be adopted and extended in their ongoing research. This book provides a good range of studies to foster further research in various software engineering domains by integrating other computer science and non-computer science domain research solutions. Software Engineering research is becoming more and more interdisciplinary in nature with ample opportunities to integrate the interdisciplinary best practices, which require a perfect blend of empirical research and evaluation research solutions to be disseminated in the literature.

The editors hope that this book will be beneficial to its intended audience and we wish them a very happy reading, happy learning and happy adoption.

Varun Gupta
Universidade da Beira Interior, Covilhã, Portugal
University of Salford, Manchester, United Kingdom

Chetna Gupta
Jaypee Institute of Information Technology, Noida, India

Editor Biographies

Varun Gupta received his PhD and M.Tech (by Research) degrees in Computer Science & Engineering from Uttarakhand Technical University, Sudowala, India, and B.Tech with Honors degree from Himachal Pradesh University, Shimla, India, respectively. He also holds an MBA degree from Pondicherry University, Pondicherry, India. He is currently working as Postdoctoral Researcher with Universidade da Beira Interior, Portugal. He is also visiting Postdoctoral Researcher at the School of Business, FHNW University of Applied Sciences and Arts Northwestern Switzerland. He was an Honorary Research Fellow of the University of Salford, Manchester, United Kingdom (2018–2021). Dr. Gupta is an Associate Editor of *IEEE Access*, Associate Editor of *International Journal of Computer Aided Engineering & Technology,* Associate Editor of the IEEE Software Blog, and Associate Editor of *Journal of Cases on Information Technology* (JCIT). His areas of interest are evidence-based software engineering, evolutionary software engineering (focusing on requirement management), business model innovation, and innovation management.

Chetna Gupta is Associate Professor in the Department of Computer Science & Engineering/Information Technology at the Jaypee Institute of Information Technology, Noida, India. She has more than 15 years of experience in research and teaching at the university level. She is currently serving as Associate Editor of the *International Journal of Computer-Aided Engineering and Technology* (Inderscience) and Associate Editor of *Journal of Cases on Information Technology* (IGI Global-USA). Furthermore, she has served as Guest Editor of many special issues published/ongoing with leading International Journals and is also editing books to be published by IGI Global and Taylor & Francis Group (CRC Press). In addition to this, she has successfully organized workshops and special sessions with Indexed International Conferences worldwide, proceedings of which were published by Springer, IEEE, Elsevier, etc. She has served as TPC in many IEEE international conferences and as a reviewer with many leading journals, including *IEEE Access*, Elsevier, Springer, etc. Her areas of interest include software engineering, search-based software engineering, distributed software engineering, risk management, automated software engineering, cloud computing, blockchain technology, and applications of machine learning and data mining.

Contributor Biographies

 **Nesar Ahmad** received his BSc (Engg) degree in Electronics & Communication Engineering from Bihar College of Engineering, Patna, India (Now NIT, Patna) in 1984. He received his MSc (Information Engineering) degree from City University, London, UK, in 1989, and his PhD degree from Indian Institute of Technology, Delhi, India, in 1993. He is currently a Professor in the Department of Computer Engineering, Aligarh Muslim University, Aligarh, India. His current research interests mainly include soft computing, web intelligence, and E-Learning. Contact him at nesar.ahmad@gmail.com

 **Tameem Ahmad** received his M.Tech. degree in Computer Science & Engineering with specialization in Software Engineering and B.Tech. degree in Computer Science & Engineering in 2013 and 2010, respectively. Currently, he is working as an Assistant Professor in the Department of Computer Engineering, Aligarh Muslim University, Aligarh, India. He has also worked in Cognizant Technology Solutions. His research areas of interest are soft computing, natural language processing, database systems, and data warehousing. Contact him at tameemahmad@gmail.com

 **Rohallah Benaboud** received his BS degree in Computer Science from Mentouri University of Constantine (Algeria) in 2002 and his MS degree in Computer Science from University of Oum El Bouaghi (Algeria) in 2005. He obtained his PhD in Computer Science from University Abdelhamid Mehri Constantine 2 (Algeria) in 2016. He is currently a member of the Distributed-Intelligent Systems Engineering (DISE) team at ReLa(CS)2 Laboratory, University of Oum El Bouaghi. He is working as Lecturer in the Department of Mathematics and Computer Science in University of Oum

El Bouaghi (Algeria). His research interests include multi-agents systems, web applications, and software engineering. Contact him at r_benaboud@yahoo.fr

Beatriz Bernárdez is an Assistant Professor at the University of Seville, Spain. Her current research focuses on empirical software engineering, requirements engineering, and the application of mindfulness in software engineering. She has the Search Inside Yourself Certification of the personal growth program (https://siyli.org/) developed in collaboration with Google. She has served as a reviewer for the IEEE *Transactions on Software Engineering* (TSE) and *Empirical Software Engineering* (EMSE) journals and has collaborated in the organization of international conferences such as ESEM'2007, SPLC'2017, and ICSE'2021. Contact her at beat@us.es

Clavers Chabi is a PhD student at the Computer Science and Engineering Department at Obafemi Awolowo University, Ile-Ife, Nigeria. He has an interest in software engineering research with a focus on mining software repositories, natural language processing, and machine learning. Contact him at claversc@yahoo.com

Kuljit Kaur Chahal is an Associate Professor in the Dept. of Computer Science, Guru Nanak Dev University, Amritsar, India. Her research interests lie in open source software, distributed systems, intelligent systems, cyber security, and computing education. She has been awarded research and travel grants by IBM Research Fund, Google, University Grants Commission (India), MHRD (India), and Guru Nanak Dev University (India). Contact her at Kuljitchahal.cse@gndu.ac.in

Francisco Chiclana received his BSc and PhD in Mathematics from the University of Granada, Spain, in 1989 and 2000, respectively. He is a Professor of Computational Intelligence and Decision Making with the School of Computer Science and Informatics, Faculty of Computing, Engineering and Media, De Montfort University, Leicester, UK. During the period 2015–2018, he was Honorary Professor in the Department of Mathematics at the University of Leicester and is currently a Visiting Scholar at the Department of Computer Science and Artificial Intelligence, University of Granada. He is an Associate Editor and Guest Editor for several ISI indexed journals. He has organized and chaired special sessions/workshops in many major international conferences in research areas such as fuzzy preference modeling, decision support systems, consensus, recommender systems, social networks, rationality/consistency, and aggregation. Clarivate Analytics has currently classed Prof. Chiclana as a Highly Cited Researcher in Computer Sciences.

Margarita Cruz is an Assistant Professor at the University of Seville, Spain, where she has been teaching databases for more than 30 years. Her current research focuses on empirical software engineering, specifically in the methodological aspects of experiment replications such as the specification of changes between replications and their reporting, not only in software engineering, but also in other research areas. Contact her at cruz@us.es

Ankush Dadwal received an M.E. degree from Waseda University, Tokyo, Japan, in 2019.

Subhajit Datta received his PhD in computer science from Florida State University, Florida. He is currently an Assistant Professor of Information Systems (Education) at the Singapore Management University (Singapore). He has more than 20 years of experience in software design, development, research, and teaching with various industrial and academic organizations in the US, India, and Singapore. In addition to numerous research publications, Subhajit is the author of two books, *Software Engineering: Concepts and Applications* (Oxford University Press, 2010) and *Metrics-Driven Enterprise Software Development* (J. Ross Publishing, 2007), which are widely used by students and practitioners. His research interests include software architecture, empirical software engineering, social computing, and network science. More details about his background and interest are available at www.dattas.net. Contact him at subhajit.datta@acm.org

Adela del-Río-Ortega is an Associate Professor and member of the ISA Research Group at University of Seville, Spain. She received her PhD with honors in 2012. Her research interests include business process management; the analysis and management of key performance indicators, both in structured processes and knowledge intensive processes; and their relationship with the decision-making process. Contact her at adeladelrio@us.es

Amador Durán is an Associate Professor of Software Engineering at the University of Seville, Spain. His current research focuses on requirements engineering, software variability, empirical software engineering, business process modeling, formal methods, and metamorphic testing. He is the author of the requirements management tool, REM, used by universities and companies in several countries. He also serves regularly as a reviewer for international journals and conferences. Contact him at amador@us.es

Bedilia Estrada-Torres is a Postdoctoral Researcher at the University of Seville, Spain, and a member of the ISA Research Group. She received her PhD with honors in 2018. She has collaborated and carried out research stays in Brazil and Estonia, and has participated in Spanish and European research projects. Her research interests include business process management; the analysis, modeling and management of process performance indicators in different scenarios, such as structured processes; variability in process families; and knowledge-intensive processes and their relationship with decision-making processes. Contact her at iestrada@us.es

Gloria Ezzat received her BSc degree in Software Engineering from the Faculty of Informatics and Computer Science, British University in Egypt in July 2019. She has been a Teaching Assistant with the Faculty of Informatics and Computer Science at the British University in Egypt since September 2019. Contact her at gloria.hanna@hotmail.com.

Yoshiaki Fukazawa received his B.E., M.E. and D.E. degrees in Electrical Engineering from Waseda University, Tokyo, Japan, in 1976, 1978, and 1986, respectively. He is now a Professor in the Department of Information and Computer Science, Waseda University, as well as the Director of Institute of Open Source Software, Waseda University. His research interests include software engineering, especially reuse of object oriented software and agent-based software.

Tamas Galli is a PhD Researcher at the Institute of Artificial Intelligence of De Montfort University, Leicester, UK. His research includes modeling the subjective manifestation of uncertainty, missing and contradicting information by means of computational intelligence methods in the field of software product quality assessments. Tamas holds an MPhil degree in Computer Science from De Montfort University, Leicester, UK, and two BEng degrees in Electrical and Electronic Engineering from the Kando Kalman College of Engineering, Budapest, Hungary. He received the following international industrial and professional certificates: Certified Scrum Master (Scrum Alliance); Chartered IT Professional Member of the British Computer Society; Sun Certified Enterprise Architect; Oracle Certified Expert: Java EE Web Component Developer; Oracle Certified Professional: Java Programmer; Sun Certified Web Component Developer; Sun Certified Business Component Developer; Sun Certified Java Programmer; and Oracle PL/SQL Certified Developer Associate. Tamas pursues his research part-time while working as a software engineer full-time. Contact him at tamas.galli@bcs.org

Ishaya Gambo is a faculty member at the Computer Science and Engineering Department at Obafemi Awolowo University, Ile-Ife, Nigeria. He was an exchange student and staff at the University of Eastern Finland in 2011 and 2015, respectively. His research is in the area of software engineering, particularly in requirements engineering, mining software repositories, software testing, and software architecture. He appreciates applying his research in the healthcare domain. The emphasis of his research is on users' and developers' perspective of software systems. Ishaya has experience of applied projects and has a clear history of the full research life cycle, including academic publishing in journals and extensive presentation at conferences. Currently, he is a Postdoctoral Research Fellow at the Institute of Computer Science, University of Tartu, Estonia. Contact him at ipgambo@oauife.edu.ng and ishaya.gambo@ut.ee

Shilpa Mariam George is a student at Oakland University (Rochester, Michigan) pursuing her MS in Software Engineering & Information Technology. She also works as a Software Developer for an IT company. She received her undergraduate degree from Rajagiri School of Engineering & Technology, India. She has prior experience working with legacy software as a developer. Contact her at kim2@oakland.edu

Abeer Hamdy received her BSc, MSc and PhD degrees in Electronics and Electrical Communications from the Faculty of Engineering, Cairo University, Egypt, in 1992, 1998, and 2003, respectively. She has been an Associate Professor with the Faculty of Informatics and Computer Science (ICS) at the British University in Egypt since 2009. Before that, she was a research scientist with the Electronics Research Institute in Egypt. She was awarded two fellowships to conduct post-doctoral research at the University of Connecticut (2005–2006) and University of Central Florida (2007–2008), USA. Her current research interests include: software design, software testing, software maintenance, and search based software engineering and computational intelligence. Contact her at abeer.hamdy@bue.edu.eg.

Rhoda Ikono is a Lecturer of Computer Science and has a PhD in Computer Science from Obafemi Awolowo University, Ile-Ife, Nigeria. Her research interest is in information systems, software engineering, health informatics, and software product usability. Contact her at rudo@oauife.edu.ng and rhoda_u@yahoo.com

Amanpreet Kaur is a PhD student in Computer Science at the Guru Nanak Dev University, Amritsar, India. She received her Master of Computer Applications degree from the same university in 2016. Her research interests include educational data mining, computational thinking, and programming performance analysis.

Abu Huzaifa Khan received his B.Tech. degree in Computer Engineering from Zakir Husain College of Engineering and Technology, Aligarh Muslim University, Aligarh, India, in 2018. Currently, he is working as a Project Engineer in Wipro Technologies, Pune, India. Contact him at huzaifakhan30@gmail.com

Dae-Kyoo Kim is a Professor in the Department of Computer Science and Engineering at Oakland University (Rochester, Michigan). He received his PhD in computer science from Colorado State University in 2004. He also worked as a technical specialist at the NASA Ames Research Center in 2002. His research interests include software engineering, software security, and data modeling in IoT and smart grids.

Selma Mammeri is a PhD student at the University of Tebessa (Algeria). She is also a principal engineer in computer science in the direction of education and a member of the LAMIS laboratory (artificial intelligence and its applications team). She obtained her engineering degree on computer science in 2010 from the University of Oum El Bouaghi, and she started her work in the direction of education as a state engineer in computer science in 2011. In 2019, she obtained her master's degree in computer science from the University of Oum El Bouaghi where she was interested in the field of artificial intelligence, more specifically multi-agent systems and normative multi-agent systems. Contact her at mammeri.sel@gmail.com

Toufik Marir is a Senior Lecturer at the Department of Mathematics and Computer Sciences, University of Oum El Bouaghi (Algeria). He received his engineering degree in computer science from the University of Oum El Bouaghi (Algeria) in 2006, and MS degree from the University of Khenchela (Algeria) in 2009. In 2015, he obtained his PhD degree from the University of Annaba (Algeria). Then, he received his habilitation from the University of Oum El Bouaghi (Algeria) in 2019. Currently, he is a member of the Distributed-Intelligent Systems Engineering (DISE) team at ReLa(CS)2 Laboratory, University of Oum El Bouaghi. His main research interests include multi-agent systems, agent-oriented software engineering, quality assurance, and software engineering. Contact him at marir.toufik@yahoo.fr

Alberto Martin-Lopez is a PhD candidate at the Applied Software Engineering research group (ISA, www.isa.us.es), University of Seville, Spain, where he received his MSc degree. His current research interests focus on automated software testing and service-oriented architectures. Contact him at amarlop@us.es

Masashi Mizoguchi is a Researcher at Hitachi, Ltd, Research & Development Group. He received his MS degree in Engineering Science from Osaka University, Japan, in 2017. His research interests are centered on software engineering for embedded control systems.

Theresa Omodunbi currently works at the Department of Computer Science & Engineering, Obafemi Awolowo University, Ile-Ife, Nigeria. She researches computational linguistics, semantics, and biomedical engineering, and health informatics. Her most recent publication is "Trends in Multi-document Summarization System Methods" in the *International Journal of Computer Applications* (July 2014).

José A. Parejo received his PhD in 2013 from the University of Seville, Spain. During the review process of this article, he became a tenured Associate Professor of the Department of Computer Languages and Systems at the same university. His research interests include empirical software engineering, meta-heuristic optimization and simulation, RESTful web services, and the combination of all of them. He is the author of the EXEMPLAR platform, a repository for experiment replication. Contact him at japarejo@us.es

Lawrence Peters has more than 50 years of experience in the field of software engineering as a programmer, systems analyst, consultant, and educator. He has worked in the aerospace, manufacturing, insurance, banking, telecommunications, and defense industries in the United States and Canada. He is the author of *Getting Results from Software Development Teams* (Microsoft Press, 2008), and other books and several papers on software project management. He holds a BS in Physics, an MS in Engineering, and a PhD in software project management. He currently teaches software project management remotely at Universidad Politecnica de Madrid, Spain, and Iowa State University, Ames, Iowa, USA.

Manuel Resinas is an Associate Professor at the University of Seville, Spain, and a member of the ISA Research Group. His current research lines include analysis and management of service level agreements, business process management, process performance analytics, and cloud-based enterprise systems. Previously, he worked on automated negotiation of service level agreements. Contact him at resinas@us.es

Antonio Ruiz-Cortés is a Professor of Computer Science at the University of Seville, Spain, and elected member of Academia Europæa. He leads the Unit of Excellence of the I3US Institute and the ISA research group. His current research focuses on service-oriented computing, business process management, experimental software engineering, testing, and software product lines. He is the recipient of the MIP Award of SPLC'2017 and VaMoS'2020. He is an associate editor of *Springer Computing*. Contact him at aruiz@us.es

Munish Saini is an Assistant Professor in the Department of Computer Engineering and Technology, Guru Nanak Dev University, Amritsar, India. He studied the evolutionary behavior of Open Source Software Projects during his PhD research. His research interests include open source software, software evolution, and mining software repositories. He has presented his research at reputed international and national forums.

Ana B. Sánchez received her PhD in software engineering (with honors) from the University of Seville, Spain, in 2016. She currently works as an Assistant Lecturer at the Department of Languages and Computer Systems at the University of Seville since 2018. She has been a member of the Applied Software Engineering Research Group since 2012. Her research interests are mainly focused on software testing. She has also participated in the organization of international and national conferences and in the review of several journals. Contact her at anabsanchez@us.es

Mohammad Saqib received B.Tech. degree in Computer Science & Engineering in 2018. Currently, he is working as a Research & Development Engineer in Tejas Networks Ltd., Gurgaon. He is mainly involved in the designing, development, and maintenance for Gigabit Passive Optical Networks (GPON) OLT's software. Contact him at msaqib4203@gmail.com

Sergio Segura is an Associate Professor of software engineering at the University of Seville, Spain. He is a member of the Applied Software Engineering research group, where he leads the research lines on software testing and search-based software engineering. His current research interests include test automation and AI-driven software engineering. Contact him at sergiosegura@us.es.

Francois Siewe is a Reader in Computer Science and Head of the Software Technology Research Laboratory (STRL) in the School of Computer Science and Informatics of the Faculty of Computing, Engineering and Media at De Montfort University (DMU), Leicester, UK. He received a BSc in Mathematics and Computer Science, and an MSc and a "Doctorat de Troisieme Cycle" in Computer Science from the University of Yaounde I in Cameroon in 1990, 1991, and 1997, respectively. Then in 2005, he received a PhD in Computer Science from De Montfort University in the UK. Before joining DMU, he was a lecturer and visiting researcher in the Institute of Technology of Lens at the University of Artois in Lens, France. Prior to this, he was a fellow at the United Nation University/International Institute for Software Technology (UNU/IIST) in Macau, China, and a lecturer in the Department of Mathematics and Computer Science at the University of Dschang in Cameroon. His research interests include software engineering, formal methods, cyber security, context-aware and pervasive computing, and the Internet of Things (IoT).

Naohiko Tsuda received his PhD in Computer Science and Engineering from Waseda University, Japan, in 2020. He is now a Junior Researcher at the Global Software Engineering Laboratory, Waseda University, Japan. His research interests include software quality evaluation and software data analysis.

Hironori Washizaki is a Professor and the Associate Dean of the Research Promotion Division at Waseda University in Tokyo, Japan, and a Visiting Professor at the National Institute of Informatics. He also works in industry as Outside Directors of SYSTEM INFORMATION and eXmotion. He received his PhD in information and computer science from Waseda University in 2003. Since 2017, he has been the lead on a large-scale grant at MEXT called enPiT-Pro SmartSE, which encompasses IoT, AI, software engineering, and business. He serves as Chair of the IEEE CS Professional and Educational Activities Board (PEAB) Engineering Discipline Committee, Associate Editor of *IEEE Transactions on Emerging Topics in Computing* (TETC), Steering Committee Member of the IEEE Conference on Software Engineering Education and Training (CSEE&T), Advisory Committee Member of the IEEE CS flagship conference COMPSAC, Convener of ISO/IEC/JTC1 SC7/WG20, and Steering Committee Member of Asia-Pacific Software Engineering Conference (APSEC).

Simon Yange holds a PhD degree in Computer Science from Obafemi Awolowo University, Ile-Ife, Nigeria. His research interest is in the areas of data science & engineering, software engineering, and information systems. He is currently an academic staff in the Department of Mathematics/Statistics/Computer Science, University of Agriculture, Makurdi, Nigeria. Contact him at lordesty2k7@ymail.com

Saradha Yasasvi Yeduruvada is a graduate student mastering in the Department of Computer Science and Engineering at Oakland University (Rochester, Michigan). She received a B.Tech. degree in computer science from JNTU University (India) in 2017. She also worked as a Salesforce Developer for a year in India and left her job to pursue her master's degree. Her research interests include advancement in IoT, software engineering, information security, and cloud computing.

Kentaro Yoshimura is a Senior Researcher at Hitachi, Ltd, Research & Development Group. He received his PhD degree in Information Science and Technology from Osaka University, Japan, in 2009. His research interests are centered on software engineering for embedded control systems.

Chapter 1

Performance of Execution Tracing with Aspect-Oriented and Conventional Approaches

Tamas Galli[a], Francisco Chiclana[b], and Francois Siewe[a]

[a]*De Montfort University, Leicester, UK*
[b]*De Montfort University, Leicester, UK, and University of Granada, Granada, Spain*

Contents

1.1 Introduction 2
1.2 Research Design 4
    1.2.1 Definition of the Research Variables 6
        1.2.1.1 Independent Variables...6
        1.2.1.2 Extraneous Variables...7
        1.2.1.3 Dependent Variables..7
1.3 Measurements and Evaluation..8
    1.3.1 Parallel Measurements between the Test Applications with
        Aspect-Oriented and Conventional Execution Tracing....................8
        1.3.1.1 Parallel Measurements with Java 1.8 and Aspect J 1.8 8
        1.3.1.2 Parallel Measurements with Java 1.9 and Aspect J 1.9.......12
    1.3.2 Comparison of the Runtimes with Switched On and Switched
        Off Execution Tracing 16

1.3.2.1 Runtime Ratios of Conventional Test Application
with Switched On and Switched Off Execution Tracing... 18
1.3.2.2 Runtime Ratios of Aspect-Oriented Test Application
with Switched On and Switched Off Execution Tracing...22
1.3.3 Comparison of the Runtimes of All Measurements with Both
Test Applications 27
1.4 Reliability and Validity 29
1.5 Conclusion..29
References 30

Appendix 1.A Normality Check of the Measurement Data..............................33

1.1 Introduction

Execution tracing assists software developers to identify and localize errors in software applications; logging rather supports the activities of system administrators and software maintainers to check the state of software systems [1, 2]. Both tools help to carry out analysis and the literature often uses the two terms as synonyms, we also follow this convention. Execution tracing and logging gain more importance in the domain of distributed programming, multithreading, real-time control applications, and embedded applications without user interface, where using a debugger is not necessarily the proper solution [3].

Execution tracing quality significantly influences the maintenance costs through its impact on the error analysis [1, 2, 4]. If the execution tracing mechanism is nonexistent, ill-designed, or of poor quality in deployed systems, then aspect-orientation offers a technique by which the missing or deficient execution tracing mechanism can be added, corrected, or its quality, with regard to accuracy and consistency, improved with low development effort.

Aspect-oriented programming facilitates the localization of functionally identical code, which is scattered across the application, and the insertion of this code, either at compilation time or at run time, into places that are identified by rules [5]. The functionally identical code segments scattered across the application are called "cross-cutting concerns". The cross-cutting concerns are collected by "advices". The modules containing the advices form the "aspects". The rules determining where to add the advices are named "pointcuts", while the concrete places the pointcuts select in the application are designated as "join points" in the aspect-oriented terminology [5, 6].

Execution tracing typically results in cross-cutting concerns. The conventional object-oriented or procedural programming approach implicates manual method call insertions to implement the execution tracing functionality in the application. In a robust system it is a monotonous, effort-intensive, and error-prone activity that usually results in inconsistent code coverage [7]. In contrast, the aspect-oriented programming approach applies automatic code insertion to the join points to achieve the same goal that ensures consistency [5, 7]. The most popular programming languages, C from the procedural domain, Java, C++, and C# from the object-oriented domain make possible to use the aspect-oriented approach for inserting trace code or for carrying out significantly more complex activities [6, 8–14]. Moreover, aspect-oriented programming is also available in the embedded and real-time domain [15, 16]. In addition, aspect-oriented functionality can be implemented in most programming languages by means of language preprocessors [17].

The aspect-oriented approach inverts the dependencies of the application code and the execution tracing functionality; as a consequence of which, in contrast to the object-oriented implementation, changes in the execution tracing mechanism will not implicate changes in the application code [5, 6]. Even the publications arguing the general use of aspect-orientation admit its applicability for execution tracing [18]. Exploiting the opportunities of aspect-oriented programming, the execution tracing mechanism of an application can be updated or completely replaced with ease [19, 20]. Aspect-oriented execution tracing can reduce the time spent on error analysis as it helps to avoid performing posterior, invasive modifications in the code base, i.e., adding additional execution trace points, to gain more information for identifying the source of the issues [19]. Moreover, aspect-orientation can also be used in the legacy domain [21].

Hilsdale and Hugunin in [22] carried out performance measurements on logging with early versions: AspectJ 1.1 on Java 1.4: with logging enabled, they experienced a negligible performance overhead of 3% but with disabled logging the upper bound of the slow down reached 2900% in the AspectJ variant in comparison to the hand-coded implementation. The best AspectJ logging implementation performed with a 22% performance overhead in their experiments [22]. Similar experiments were carried out by Colyer et al. in [19], where they experienced a 5% performance overhead with disabled logging but after eliminating the use of thisJointPoint they measured a performance overhead of 1% or less. Ali et al. in [23] point out contradictions in their extensive study of 3307 publications related aspect-orientation with Systematic Literature Review [24]. They examined 22 empiric studies in-depth, which compared aspect-oriented and non-aspect-oriented techniques [23]. Only 7 of the 22 studies made statements on performance but only 3 of the 7 reports are based on experimental research, 2 of which claimed positive performance impact and 1 no significant impact. In addition, the 3 performance-related experimental studies used different AOP implementations: AspectC, GlountJ, and AspectJ [23].

After the early years and its diverse fame on performance, aspect-oriented programming found its way in the most programming frameworks including Spring,. NET [11, 25], and Java EE; however, Enterprise Java Beans do not offer full support [26]. Even if, according to the majority of the publications, the aspect-oriented approach causes performance overhead, this does not neutralizes its advantages to add functionality to a code base in a simple automatic and consistent manner, which should otherwise be done in an effort-intensive, error-prone manual way. Many publications deal with the performance of aspect-oriented programming but concrete measurements and benchmark tests published are usually old using old programming platforms [27]. Consequently, the questions of aspect-oriented performance should be revisited, especially with regard to implementing execution tracing, which is currently its most frequent and simplest use case in the practice.

The chapter is structured as follows: in Section 1.2 the research methods are introduced including the definition of the research variables; in Section 1.3 the charts coming from the measurements are illustrated and execution tracing performance is evaluated; Section 1.4 focuses on the reliability and validity of the study; Section 1.5 ends with the conclusion.

1.2 Research Design

As mentioned above, performance, beside accuracy and consistency, is one of the major concerns of the software developers with regard to execution tracing. In addition, publications on the performance of the aspect-oriented implementations deviate in their conclusions regarding this quality property. Moreover, the analyzability of the software errors strongly depends on the execution tracing mechanism of the application [1, 2, 7].

Addressing the current situation, the present research aims at investigating of the performance with true experimental research [28] to test cause-effect relationship for a definite, well-documented use case: execution tracing with one aspect-oriented framework for Java: AspectJ. The experiments span multiple matching Java-AspectJ versions released in the last years: Java 1.8-AspectJ 1.8, and Java 1.9-AspectJ 1.9. AspectJ 1.9.5 is currently the latest available release issued on 28.11.2019.

The experiment contains the following software artefacts: a test application pair where both of the applications do the same activities with differences in the tracing mechanism. Both test applications were designed and implemented in the same way for the same purpose. One application implemented execution tracing in the object-oriented manner by inserting the calls of the trace methods manually. The other application applied a tracing aspect to achieve the same goal. The aspect-oriented test application implemented singleton aspect association, i.e., one instance of the aspect type existed during the test run. The thisJoinPointStaticPart language construct was used where possible instead of its dynamic counterpart; moreover, loops were used within the aspect only if they were really necessary. In both test applications, the trace messages were only constructed when execution tracing was activated. The runtime of each application was measured in msecs. Between the points of the time measurement both test applications produced the same amount of bytes when tracing was activated.

Goal of the experiment: (1) to compare the performance of the parallel running test applications; (2) to compare the runtimes of activated and deactivated

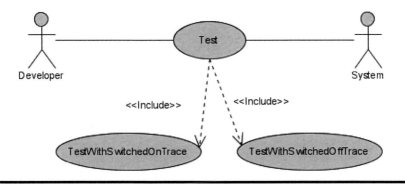

Figure 1.1   Use case diagram of the test applications.

execution tracing for each test application. Operational definitions of the input and output variables are in Section 1.2.1.

The hardware platform was a virtual machine with Intel Xeon CPU E5-2697 2.60GHz processor and 8 GB RAM.

The use case diagram Figure 1.1 shows that the test includes test cases with both switched-on and switched-off state of execution tracing.

For simulating high load in the test applications, merging string, and integer arrays are used. The merged array is returned from the method that created it; thus, it is also serialized and traced if execution tracing is switched on. In addition, the call stack is increased with recursion. A test run of each application comprises of 100.000 iterations before exiting. One measurement contains 10 test suites and each test suite contains 20 test runs to produce series of data for testing differences statistically.

The sequence diagram Figure 1.2 depicts the message flows during the test runs. Explanation of the message flows:

1. The application is launched and it creates the number of necessary threads to run the test (Number of threads in the experiments=1).
2. Message 8 launches a loop with the number of iterations shown by the variable MEASUREMENT_CYCLE (MEASUREMENT_CYCLE=100.000 in the experiments).
3. Message 8 contains a recursive call; the variable DEPTH_OF_CALL_STACK controls the call stack depth (DEPTH_OF_CALL_STACK=10 in the experiments).
4. Message 7 calls a mediator method, doSomeJob(), which implements string operations in a loop with the number of iterations shown by the variable JOB_LENGTH for ensuring latency and to simulate load. This loop is not depicted on the diagram for better readability (JOB_LENGTH=100 in the experiments).

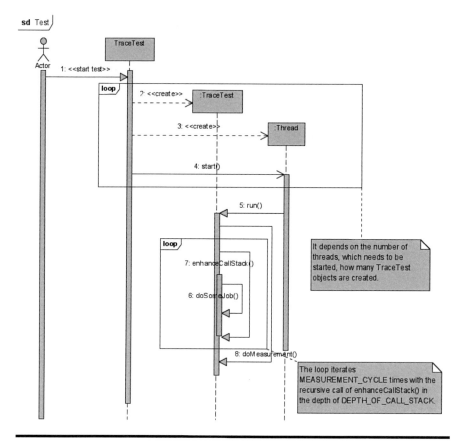

Figure 1.2 Sequence diagram of the test applications.

1.2.1 Definition of the Research Variables

1.2.1.1 Independent Variables

1. Type of execution tracing mechanism: Aspect-oriented (AOP) or conventional (CONV)

$$I_{\text{TraceType}} = \{AOP, CONV\}$$

2. State of the execution tracing: Switched on or switched off

$$I_{\text{TraceState}} = \{ON, ONREALFILE, OFF\}$$

3. Java Version: Version of the Java environment

$$I_{JavaVersion} = \{1.8, 1.9\}$$

4. ApectJ Version: Version of the AspectJ environment

$$I_{AJVersion} = \{1.8, 1.9\}$$

1.2.1.2 Extraneous Variables

1. Any other activity taking place on the target host: Activities that might have an impact on the runtime of the test applications

$$E_{OtherInfluencingActivities} = \{Anyinfluencingactivity\}$$

1.2.1.3 Dependent Variables

1. Runtime of each test application: The runtime of the test application with aspect-oriented and conventional execution tracing.

$$T_{AOP} = AOP_{Runtime}$$

$$T_{CONV} = CONV_{Runtime}$$

2. Ratio AOP/CONV: The ratio of the runtimes of the test applications with aspect-oriented and the conventional execution tracing.

$$O_{AOP/CONV} = \frac{Runtime_{AOP}}{Runtime_{CONV}}$$

3. Ratio CONV ON/CONV OFF: Ratio of the runtimes of the test application with conventional execution tracing switched on while trace output is redirected to/dev/null and switched off.

$$O_{CONVON/OFF} = \frac{Runtime_{CONV\_ON}}{Runtime_{CONV\_OFF}}$$

4. Ratio of the runtimes of the test application with conventional execution tracing switched on while trace output is stored in a real file on the file system and switched off.

$$O_{CONVON/OFF} = \frac{Runtime_{CONV\_ON\_REALFILE}}{Runtime_{CONV\_OFF}}$$

5. Ratio of the runtimes of the test application with aspect-oriented execution tracing switched on while trace output is redirected to/dev/null and switched off.

$$O_{AOPON/OFF} = \frac{\mathrm{Runtime}_{AOP\_ON}}{\mathrm{Runtime}_{AOP\_OFF}}$$

6. Ratio of the runtimes of the test application with aspect-oriented execution tracing switched on while trace output is stored in a real file on the file system and switched off.

$$O_{AOPON/OFF} = \frac{Runtime_{AOP\_ON\_REALFILE}}{Runtime_{AOP\_OFF}}$$

As the test applications run in AOP-CONV pairs, the effects of the extraneous variables are eliminated for the variable $O_{AOP/CONV}$ only. Nevertheless, due to the high number of repetitions in the measurements, the effects of the potential extraneous variables are reduced.

1.3 Measurements and Evaluation

In this section we present the measurements, the statistical tests carried out to investigate the presence of significant differences, and the evaluation of the experiments. The distributions of the data produced by the measurements are depicted for each test suite in Appendix 1.A.

1.3.1 Parallel Measurements between the Test Applications with Aspect-Oriented and Conventional Execution Tracing

The measurements introduced in this section were carried out at the same time at the same host to eliminate the impacts of extraneous variables while computing the ratios between the test runs.

1.3.1.1 Parallel Measurements with Java 1.8 and Aspect J 1.8

- ◼ $I_{TraceType} = $ AOP and CONV as the test application run at the same time
- ◼ $I_{TraceState} = $ OFF
- ◼ $I_{JavaVersion} = 1.8$
- ◼ $I_{AJVersion} = 1.8$

The measurement data collected and their distribution charts are in Appendix 1.A.1 (see Table 1.A.1, Figure 1.A.1) and in Appendix 1.A.7 (see Table 1.A.7, Figure 1.A.7).

The quantile–quantile plots show that AOP test suites 3, 4, 5 deviate from the theoretical normal distribution at the left tail while AOP test suites 1, 2, 6, 7, 8, 9, 10 deviate at the right tail. Similar deviations can be observed with the CONV test suites 1, 3 at the left tail and CONV test suites 2, 3, 8, 9, 10 at the right tail. Consequently, the statistical test needs to investigate the differences in the median between the two measurements, for which the Mann–Whitney test was used (Table 1.1). The runtime rations are depicted in Figure 1.3.

- $I_{TraceType}$ = AOP and CONV as the test application run at the same time
- $I_{TraceState}$ = ON
- $I_{JavaVersion}$ = 1.8
- $I_{AJVersion}$ = 1.8

Execution tracing was switched on in both test applications and the trace output was redirected in/dev/null. The measurement data collected and their distribution charts are in Appendix 1.A.2 (see Table 1.A.2, Figure 1.A.2) and in Appendix 1.A.8 (see Table 1.A.8, Figure 1.A.8). The quantile–quantile plots show that AOP test suites 3, 4, 5, 7, 10 deviate from the theoretical normal distribution at the left tail while AOP test suites 2, 3, 7, 9 deviate at the right tail. No deviations can be observed with the CONV test suits at the left tail but the CONV test suites 3, 5, 6 deviate from the theoretical normal distribution at the right tail. Consequently, the statistical test needs to investigate the differences in the median between the two measurements, for which the Mann–Whitney test was used (Table 1.2). The runtime rations are depicted in Figure 1.4.

- $I_{TraceType}$ = AOP and CONV as the test application run at the same time
- $I_{TraceState}$ = ON REAL FILE
- $I_{JavaVersion}$ = 1.8
- $I_{AJVersion}$ = 1.8

Execution tracing was switched on in both test applications and the trace output was stored in a real file on the file system. The measurement data collected and their distribution charts are in Appendix 1.A.3 (see Table 1.A.3, Figure 1.A.3) and in Appendix 1.A.9 (see Table 1.A.9, Figure 1.A.9). The quantile–quantile plots show that AOP test suites 2, 9 deviate from the theoretical normal distribution at the left tail while AOP test suite 3 deviates at the right tail. Similar deviations can be observed with the CONV test suites 1, 5, 7 at the left tail and the CONV test suites 4, 9 deviate from the theoretical normal distribution at the right tail. Consequently, the statistical test needs to investigate the differences in the median between the two measurements, for which the Mann–Whitney test was used (Table 1.3). The runtime rations are depicted in Figure 1.5.

Table 1.1 Mann–Whitney Test p-values for the Medians with Mean and Median Ratios per Test Suites, Parallel Runs, Java 1.8, AJ 1.8, Trace Off

	Suite 1	Suite 2	Suite 3	Suite 4	Suite 5	Suite 6	Suite 7	Suite 8	Suite 9	Suite 10
p-value	6.79E-08	6.79E-08	0.000258	1.45E-11	1.45E-11	7.89E-08	9.92E-09	1.45E-11	6.79E-08	1.45E-11
Mean ratio	1.176716	1.189654	1.193483	1.172909	1.186373	1.169318	1.158561	1.176	1.151761	1.192008
Median ratio	1.161067	1.172599	1.268788	1.161551	1.176939	1.153497	1.164981	1.172307	1.150186	1.180376

Table 1.2 Mann–Whitney Test p-values for the Medians with Mean and Median Ratios per Test Suites, Parallel Runs, Java 1.8, AJ 1.8, Trace On (/dev/null)

	Suite 1	Suite 2	Suite 3	Suite 4	Suite 5	Suite 6	Suite 7	Suite 8	Suite 9	Suite 10
p-value	0.000933	1.83E-05	0.005131	0.000119	0.000258	0.016742	9.65E-06	0.002166	0.000292	0.002915
Mean ratio	1.059076	1.070867	1.065352	1.070544	1.085757	1.049475	1.092887	1.070977	1.079785	1.080769
Median ratio	1.069102	1.075855	1.060753	1.049596	1.086463	1.058268	1.067856	1.066465	1.090995	1.077837

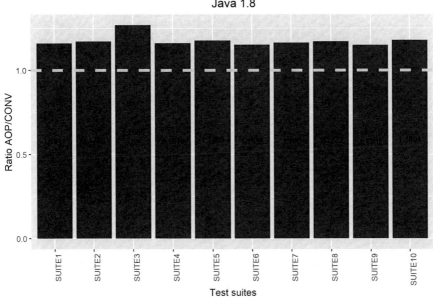

Figure 1.3 **Runtime ratio of the test applications, parallel runs, Java 1.8, AJ 1.8, trace off.**

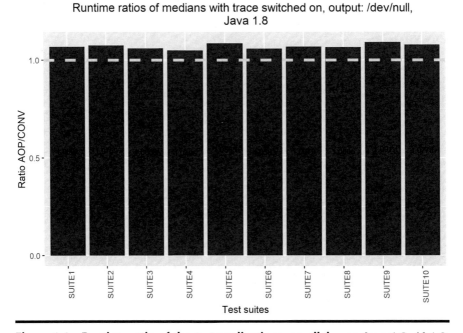

Figure 1.4 **Runtime ratio of the test applications, parallel runs, Java 1.8, AJ 1.8, trace on (/dev/null).**

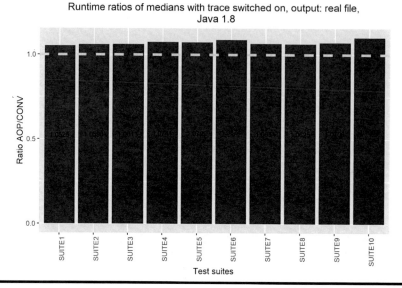

Figure 1.5   Runtime ratio of the test applications, parallel runs, Java 1.8, AJ 1.8, trace on (real file).

1.3.1.2 Parallel Measurements with Java 1.9 and AspectJ 1.9

- ■ $I_{TraceType}$ = AOP and CONV as the test application run at the same time
- ■ $I_{TraceState}$ = OFF
- ■ $I_{JavaVersion}$ = 1.9
- ■ $I_{AJVersion}$ = 1.9

The measurement data collected and their distribution charts are in Appendix 1.A.4 (see Table 1.A.4, Figure 1.A.4) and in Appendix 1.A.10 (see Table 1.A.10, Figure 1.A.10). The quantile–quantile plots show that AOP test suites 4, 5, 9 deviate from the theoretical normal distribution at the left tail while AOP test suites 2, 4, 5, 7, 10 deviate at the right tail. Similar deviations can be observed with the CONV test suite 3 at the left tail and CONV test suites 3, 10 at the right tail. Consequently, the statistical test needs to investigate the differences in the median between the two measurements, for which the Mann–Whitney test was used (Table 1.4). The runtime rations are depicted in Figure 1.6.

- ■ $I_{TraceType}$ = AOP and CONV as the test application run at the same time
- ■ $I_{TraceState}$ = ON
- ■ $I_{JavaVersion}$ = 1.9
- ■ $I_{AJVersion}$ = 1.9

Execution tracing was switched on in both test applications and the trace output was redirected in/dev/null. The measurement data collected and their distribution

Table 1.3 Mann–Whitney Test p-values for the Medians with Mean and Median Ratios per Test Suites, Parallel Runs, Java 1.8, AJ 1.8, Trace On (Real File)

	Suite 1	Suite 2	Suite 3	Suite 4	Suite 5	Suite 6	Suite 7	Suite 8	Suite 9	Suite 10
p-value	0.001435	0.002394	0.000119	0.000472	1.13E-06	0.003211	0.000595	0.003483	0.000836	4.12E-06
Mean ratio	1.053317	1.043239	1.069095	1.057292	1.075002	1.060962	1.052232	1.052818	1.055775	1.082992
Median ratio	1.052506	1.059605	1.061346	1.074103	1.072876	1.087374	1.065073	1.062538	1.07063	1.10034

Runtime ratios of medians with trace switched off,
Java 1.9

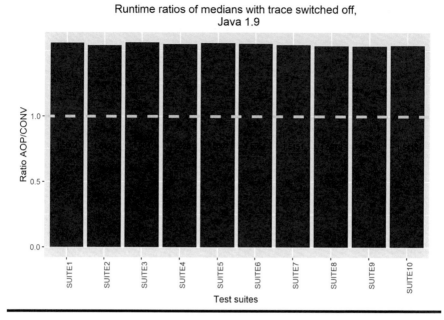

Figure 1.6 Runtime ratio of the test applications, Java 1.9, AJ 1.9, trace off.

charts are in Appendix 1.A.5 (see Table 1.A.5, Figure 1.A.5), and in Appendix 1.A.11 (see Table 1.A.11, Figure 1.A.11). The quantile–quantile plots show that AOP test suite 10 deviates from the theoretical normal distribution at the left tail while AOP test suites 2, 6, 7, 9, 10 deviate at the right tail. Similar deviations can be observed with the CONV test suites 1, 3, 6 at the left tail and the CONV test suites 2, 8, 9, 10 deviate from the theoretical normal distribution at the right tail. Consequently, the statistical test needs to investigate the differences in the median between the two measurements, for which the Mann–Whitney test was used (Table 1.5). The runtime rations are depicted in Figure 1.7.

- ■ $I_{TraceType}$ = AOP and CONV as the test application run at the same time
- ■ $I_{TraceState}$ = ON REAL FILE
- ■ $I_{JavaVersion}$ = 1.9
- ■ $I_{AJVersion}$ = 1.9

Execution tracing was switched on in both test applications and the trace output was stored in a real file on the file system. The measurement data collected and their distribution charts are in Appendix 1.A.6 (see Table 1.A.6, Figure 1.A.6) and in Appendix 1.A.12 (see Table 1.A.12, Figure 1.A.12). The quantile–quantile plots show that AOP test suites 1, 5, 7, 8, 9 deviate from the theoretical normal distribution at the left tail while AOP test suites 1, 3, 7, 9 deviate at the right tail.

Table 1.4 Mann–Whitney Test p-values for the Medians with Mean and Median Ratios per Test Suites, Parallel Runs, Java 1.9, AJ 1.9, Trace Off

	Suite 1	Suite 2	Suite 3	Suite 4	Suite 5	Suite 6	Suite 7	Suite 8	Suite 9	Suite 10
p-value	1.45E-11	1.45E-11	1.45E-11	1.45E-11	1.45E-11	6.78E-08	1.45E-11	6.79E-08	1.45E-11	6.79E-08
Mean ratio	1.549438	1.536558	1.542429	1.533721	1.564503	1.558011	1.549189	1.533397	1.551533	1.53376
Median ratio	1.560213	1.542441	1.564017	1.552623	1.560204	1.555363	1.545072	1.537941	1.535937	1.541483

Table 1.5 Mann–Whitney Test p-values for the Medians with Mean and Median Ratios per Test Suites, Parallel Runs, Java 1.9, AJ 1.9, Trace On (/dev/null)

	Suite 1	Suite 2	Suite 3	Suite 4	Suite 5	Suite 6	Suite 7	Suite 8	Suite 9	Suite 10
p-value	6.79E-08	1.45E-11	1.45E-11	1.45E-11	1.45E-11	1.45E-11	1.45E-11	1.45E-11	4.35E-10	1.45E-11
Mean ratio	1.210817	1.221985	1.223094	1.218294	1.203636	1.219994	1.217966	1.220482	1.209873	1.215399
Median ratio	1.203633	1.213829	1.199321	1.223309	1.214949	1.229891	1.204611	1.201948	1.193384	1.204965

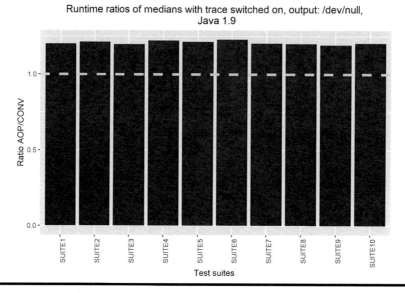

Figure 1.7 **Runtime ratio of the test applications, parallel runs, Java 1.9, AJ 1.9, trace on (/dev/null).**

Similar deviations can be observed with the CONV test suites 5, 9 at the left tail and the CONV test suites 2, 6, 8, 10 deviate from the theoretical normal distribution at the right tail. Consequently, the statistical test needs to investigate the differences in the median between the two measurements, for which the Mann–Whitney test was used (Table 1.6). The runtime rations are depicted in Figure 1.8.

1.3.2 Comparison of the Runtimes with Switched On and Switched Off Execution Tracing

The measurements used in this section were not carried out at the same time. Consequently, the runtime ratios computed were exposed to the impacts of extraneous variables while the tests run. Nevertheless, each measurement contains high number of repetitions as introduced in Section 1.2 to reduce these effects.

The computations use the same data for the comparison as in the previous Section 1.3.1. However, the view of the investigation targets not at the change of the variable trace type ($I_{TraceType} = \{AOP, CONV\}$) but at the variable trace state ($I_{TraceState} = \{ON, ON REAL FILE, OFF\}$). In front of each chart we list the corresponding input variables.

Table 1.6 Mann–Whitney Test p-values for the Medians with Mean and Median Ratios per Test Suites, Parallel Runs, Java 1.9, AJ 1.9, Trace On (Real File)

	Suite 1	Suite 2	Suite 3	Suite 4	Suite 5	Suite 6	Suite 7	Suite 8	Suite 9	Suite 10
p-value	2.90E-11	5.41E-09	1.02E-10	2.90E-11	9.72E-10	1.02E-10	2.90E-11	1.45E-11	1.45E-11	2.90E-11
Mean ratio	1.152475	1.127715	1.154458	1.144766	1.163146	1.147582	1.149267	1.155305	1.150014	1.149023
Median ratio	1.161545	1.142994	1.164729	1.134599	1.168637	1.152347	1.151139	1.166117	1.134792	1.164991

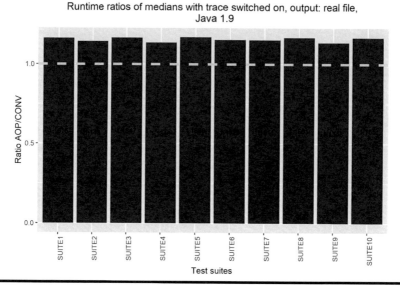

Runtime ratios of medians with trace switched on, output: real file, Java 1.9

Figure 1.8 Runtime ratio of the test applications, parallel runs, Java 1.9, AJ 1.9, trace on (real file).

1.3.2.1 Runtime Ratios of Conventional Test Application with Switched On and Switched Off Execution Tracing

- $I_{TraceType} = CONV$
- $I_{TraceState} = ON$ and OFF
- $I_{JavaVersion} = 1.8$
- $I_{AJVersion} = 1.8$

The ratios depicted in Figure 1.9 show the performance impact of the trace message construction including serializing method input parameters and return values. Trace file creation was ruled out of the comparison as the trace output was redirected to/dev/null. The p-values of the statistical tests are listed in Table 1.7.

- $I_{TraceType} = CONV$
- $I_{TraceState} = ON$ REAL FILE and OFF
- $I_{JavaVersion} = 1.8$
- $I_{AJVersion} = 1.8$

The ratios depicted in Figure 1.10 show the performance impact of the trace message construction including serializing method input parameters, return values, and writing the trace output into a trace file. One run of the test applications produces approx. 2.8 GB trace data per application. The p-values of the statistical tests are listed in Table 1.8.

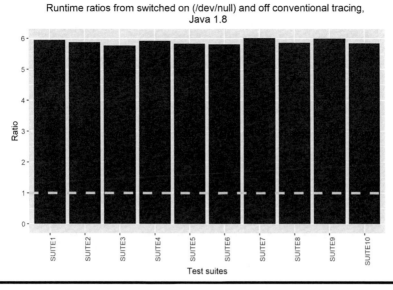

Figure 1.9 Runtime ratio of the test applications, trace type conventional, Java 1.8, trace on (/dev/null) and off.

- $I_{TraceType}$ = CONV
- $I_{TraceState}$ = ON and OFF
- $I_{JavaVersion}$ = 1.9
- $I_{AJVersion}$ = 1.9

The ratios depicted in Figure 1.11 show the performance impact of the trace message construction including serializing method input parameters and return values. Trace file creation was ruled out of the comparison as the trace output was redirected to/dev/null. The p-values of the statistical tests are listed in Table 1.9.

- $I_{TraceType}$ = CONV
- $I_{TraceState}$ = ON REAL FILE and OFF
- $I_{JavaVersion}$ = 1.9
- $I_{AJVersion}$ = 1.9

The ratios depicted in Figure 1.12 show the performance impact of the trace message construction including serializing method input parameters, return values, and writing the trace output into a trace file. One run of the test applications produces approx. 2.8 GB trace data per application. The p-values of the statistical tests are listed in Table 1.10.

Table 1.7 Mann–Whitney Test p-values for the Medians with Mean and Median Ratios per Test Suites, Trace Type Conventional, Java 1.8, Trace On (/dev/null) and Off

	Suite 1	Suite 2	Suite 3	Suite 4	Suite 5	Suite 6	Suite 7	Suite 8	Suite 9	Suite 10
p-value	6.79E-08	6.79E-08	1.45E-11	1.45E-11	1.45E-11	6.79E-08	1.45E-11	1.45E-11	6.79E-08	1.45E-11
Mean ratio	5.967495	5.917304	5.294657	5.888579	5.834671	5.822949	5.969259	5.863678	5.936071	5.844406
Median ratio	5.952594	5.880222	5.768236	5.919887	5.827043	5.811281	6.009418	5.862483	5.991051	5.840696

Table 1.8 Mann–Whitney Test p-values for the Medians with Mean and Median Ratios per Test Suites, Trace Type Conventional, Java 1.8, Trace On (Real File) and Off

	Suite 1	Suite 2	Suite 3	Suite 4	Suite 5	Suite 6	Suite 7	Suite 8	Suite 9	Suite 10
p-value	6.79E-08	6.79E-08	1.45E-11	1.45E-11	1.45E-11	6.79E-08	1.45E-11	1.45E-11	6.79E-08	1.45E-11
Mean ratio	8.251128	8.298515	7.302974	8.18246	8.097213	8.388316	8.11288	8.395151	8.242461	8.282741
Median ratio	8.247116	8.212798	7.94661	8.188654	8.145205	8.14977	8.217889	8.314315	8.325266	8.262757

Table 1.9 Mann–Whitney Test p-values for the Medians with Mean and Median Ratios per Test Suites, Trace Type Conventional, Java 1.9, Trace On (/dev/null) and Off

	Suite 1	Suite 2	Suite 3	Suite 4	Suite 5	Suite 6	Suite 7	Suite 8	Suite 9	Suite 10
p-value	6.79E-08	1.45E-11	1.45E-11	1.45E-11	1.45E-11	6.78E-08	1.45E-11	6.79E-08	1.45E-11	6.79E-08
Mean ratio	6.979438	6.867573	6.867283	6.874555	6.965891	6.89179	6.86507	6.884597	7.185274	6.87195
Median ratio	7.029971	6.948272	6.986301	6.88027	6.946895	6.847967	6.958917	6.940671	7.136755	6.913806

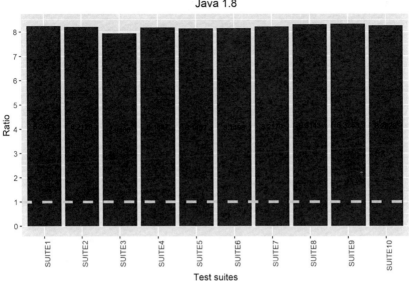

Figure 1.10 Runtime ratio of the test applications, trace type conventional, Java 1.8, trace on (real file) and off.

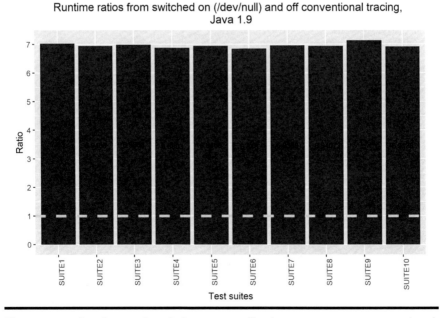

Figure 1.11 Runtime ratio of the test applications, trace type conventional, Java 1.9, trace on (/dev/null) and off.

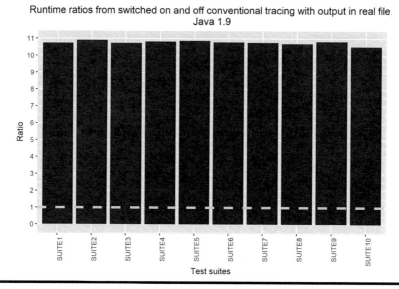

Figure 1.12 Runtime ratio of the test applications, trace type conventional, Java 1.9, trace on (real file) and off.

1.3.2.2 Runtime Ratios of Aspect-Oriented Test Application with Switched On and Switched Off Execution Tracing

- $I_{TraceType} = AOP$
- $I_{TraceState} = ON\ and\ OFF$
- $I_{JavaVersion} = 1.8$
- $I_{AJVersion} = 1.8$

The ratios depicted in Figure 1.13 show the performance impact of the trace message construction including serializing method input parameters and return values. Trace file creation was ruled out of the comparison as the trace output was redirected to/dev/null. The p-values of the statistical tests are listed in Table 1.11.

- $I_{TraceType} = AOP$
- $I_{TraceState} = ON\ REAL\ FILE\ and\ OFF$
- $I_{JavaVersion} = 1.8$
- $I_{AJVersion} = 1.8$

The ratios depicted in Figure 1.14 show the performance impact of the trace message construction including serializing method input parameters, return values, and writing the trace output into a trace file. One run of the test applications

Table 1.10 Mann–Whitney Test p-values for the Medians with Mean and Median Ratios per Test Suites, Trace Type Conventional, Java 1.9, Trace On (Real File) and Off

	Suite 1	Suite 2	Suite 3	Suite 4	Suite 5	Suite 6	Suite 7	Suite 8	Suite 9	Suite 10
p-value	1.45E-11	1.45E-11	1.45E-11	1.45E-11	1.45E-11	6.78E-08	1.45E-11	6.79E-08	1.45E-11	6.79E-08
Mean ratio	10.81557	10.85988	10.66393	10.7041	10.84204	10.82493	10.67403	10.65194	10.7508	10.5679
Median ratio	10.74297	10.91361	10.74382	10.82738	10.8887	10.79985	10.77863	10.71335	10.83527	10.52922

Table 1.11 Mann–Whitney Test p-values for the Medians with Mean and Median Ratios per Test Suites, Trace Type AOP, AJ 1.8, Trace On (/dev/null) and Off

	Suite 1	Suite 2	Suite 3	Suite 4	Suite 5	Suite 6	Suite 7	Suite 8	Suite 9	Suite 10
p-value	1.45E-11	1.45E-11	1.45E-11	1.45E-11	1.45E-11	1.45E-11	1.45E-11	1.45E-11	1.45E-11	1.45E-11
Mean ratio	5.370906	5.326461	4.726232	5.37466	5.339836	5.22616	5.630887	5.340017	5.565112	5.299001
Median ratio	5.481105	5.395081	4.822455	5.349305	5.379095	5.33152	5.50841	5.333189	5.68274	5.333315

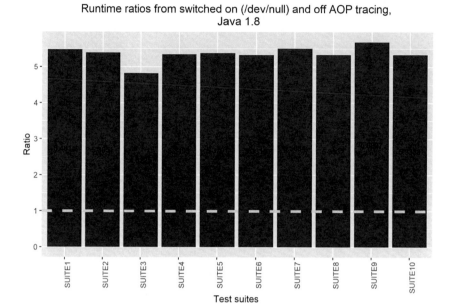

Figure 1.13 **Runtime ratio of the test applications, trace type AOP, AJ 1.8, trace on (/dev/null) and off.**

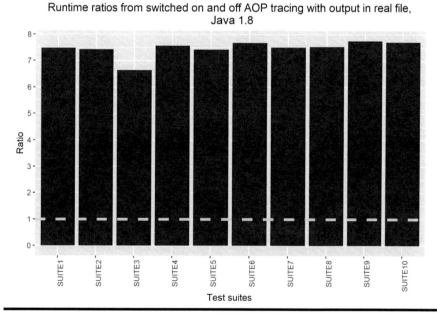

Figure 1.14 **Runtime ratio of the test applications, trace type AOP, AJ 1.8, trace on (real file) and off.**

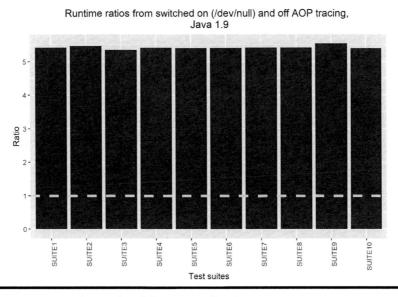

Figure 1.15 **Runtime ratio of the test applications, trace type AOP, AJ 1.9, trace on (/dev/null) and off.**

produces approx. 2.8 GB trace data per application. The p-values of the statistical tests are listed in Table 1.12.

- $I_{TraceType}$ = AOP
- $I_{TraceState}$ = ON and OFF
- $I_{JavaVersion}$ = 1.9
- $I_{AJVersion}$ = 1.9

The ratios depicted in Figure 1.15 show the performance impact of the trace message construction including serializing method input parameters and return values. Trace file creation was ruled out of the comparison as the trace output was redirected to/dev/null. The p-values of the statistical tests are listed in Table 1.13.

- $I_{TraceType}$ = AOP
- $I_{TraceState}$ = ON REAL FILE and OFF
- $I_{JavaVersion}$ = 1.9
- $I_{AJVersion}$ = 1.9

The ratios depicted in Figure 1.16 show the performance impact of the trace message construction including serializing method input parameters, return values, and writing the trace output into a trace file. One run of the test applications produces approx. 2.8 GB trace data per application. The p-values of the statistical tests are listed in Table 1.14.

Table 1.12 Mann–Whitney Test p-values for the Medians with Mean and Median Ratios per Test Suites, Trace Type AOP, AJ 1.8, Trace On (Real File) and Off

	Suite 1	Suite 2	Suite 3	Suite 4	Suite 5	Suite 6	Suite 7	Suite 8	Suite 9	Suite 10
p-value	1.45E-11	1.45E-11	1.45E-11	1.45E-11	1.45E-11	1.45E-11	1.45E-11	1.45E-11	1.45E-11	1.45E-11
Mean ratio	7.385852	7.277188	6.541838	7.37589	7.337085	7.611005	7.368309	7.515787	7.555546	7.525234
Median ratio	7.476002	7.421398	6.647374	7.572173	7.425019	7.682589	7.513131	7.535801	7.749425	7.702495

Table 1.13 Mann–Whitney Test p-values for the Medians with Mean and Median Ratios per Test Suites, Trace Type AOP, AJ 1.9, Trace On (/dev/null) and Off

	Suite 1	Suite 2	Suite 3	Suite 4	Suite 5	Suite 6	Suite 7	Suite 8	Suite 9	Suite 10
p-value	1.45E-11	1.45E-11	1.45E-11	1.45E-11	1.45E-11	1.45E-11	1.45E-11	1.45E-11	1.45E-11	1.45E-11
Mean ratio	5.45412	5.461602	5.445522	5.460726	5.359145	5.396585	5.397289	5.479683	5.603018	5.445545
Median ratio	5.423302	5.467969	5.357241	5.420951	5.409628	5.414975	5.425501	5.424348	5.545079	5.404468

Table 1.14 Mann–Whitney Test p-values for the Medians with Mean and Median Ratios per Test Suites, Trace Type AOP, AJ 1.9, Trace On (Real File) and Off

	Suite 1	Suite 2	Suite 3	Suite 4	Suite 5	Suite 6	Suite 7	Suite 8	Suite 9	Suite 10
p-value	1.45E-11	1.45E-11	1.45E-11	1.45E-11	1.45E-11	1.45E-11	1.45E-11	1.45E-11	1.45E-11	1.45E-11
Mean ratio	8.044643	7.970315	7.981604	7.989514	8.060622	7.973305	7.918535	8.025471	7.968616	7.916984
Median ratio	7.997912	8.087307	8.000966	7.912244	8.155946	8.00146	8.030498	8.123209	8.005394	7.957557

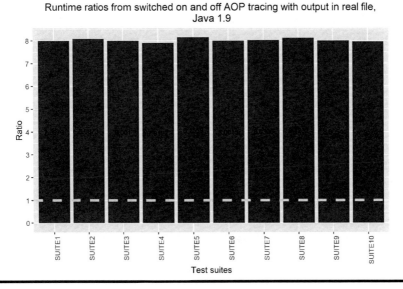

Figure 1.16   Runtime ratio of the test applications, trace type AOP, AJ 1.9, trace on (real file) and off.

1.3.3 Comparison of the Runtimes of All Measurements with Both Test Applications

This section comprises of the differences between the versions Java 1.8 and 1.9; moreover, between AspectJ 1.8 and 1.9 with regard to all other input variables.

- $I_{TraceType} = AOP \, and \, CONV$
- $I_{TraceState} = ON \, and \, OFF$
- $I_{JavaVersion} = 1.8 \, and \, 1.9$
- $I_{AJVersion} = 1.8 \, and \, 1.9$

The chart below depicts the Q1 and Q3 quartiles in the boxplots, i.e., 50% of the data fall in the range illustrated by the box. In addition, the median is depicted by the middle bold line in each boxplot while the outliers are shown with the dots over the whiskers.

The distance of the whiskers (W) from the Q1 and Q3 quartiles is computed by 1.5 times the interquartile range (IQR), i.e., the difference between Q3 and Q1. If the exact *W* value computed is not present in the data set, then the whiskers are placed to the closest value toward the median.

$$W_1 = Q1 - 1.5 * IRQ$$

$$W_2 = Q3 + 1.5 * IRQ$$

Table 1.15 Mann–Whitney Test p-values for the Runtime Medians for Different Versions, with the Combinations of the Input Variables

	CONV with Trace ON	CONV with Trace ON REAL FILE	CONV with Trace OFF	AOP with Trace ON	AOP with Trace ON REAL FILE	AOP with Trace OFF
p-value	4.83E-67	4.83E-67	4.82E-67	1.63E-66	9.21E-67	4.83E-67
Median 1.8	46765	65201	7940	49998.5	69521	9257
Median 1.9	32230	50015.5	4631.5	38986	57610.5	7186.5

The medians of the versions 1.8 and 1.9 show that the Java and AspectJ versions have a strong, statistically significant ($p < 0.001$) different impact on the performance as illustrated in Table 1.15. In each measurement, test applications performed better with the higher versions as demonstrated in Figure 1.17. Moreover, the investigation of the boxplot and the ratio bar charts show that even if the ratios between switched on and switched off execution tracing increased while changing from Java 1.8 and AspectJ 1.8 to Java 1.9 and AspectJ 1.9, the ratio increase was a consequence that the test applications achieved a performance betterment with switched off tracing to a greater extent than with switched on tracing.

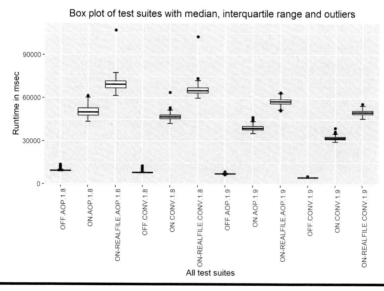

Figure 1.17 Boxplot of the runtimes of all measurements data per input variable combinations.

1.4 Reliability and Validity

The reliability of the study lies on strong statistical evidence, the level of significance of the statements made exceeds the usual p=0.05 significance value extensively. To eliminate disturbing outside effects during the experiments, so-called extraneous variables, the test applications were run in pairs with aspect-oriented and conventional execution tracing. Consequently, the ratios computed from these test runs are not affected by the covariates.

Nevertheless, the statements made are valid for the test applications under investigation and for their execution tracing mechanisms, which implemented a pressing tracing policy, in which all arguments and return values of all methods were traced. For the generalizability of the findings, the examination of open-source, real-life applications would be necessary.

1.5 Conclusion

We used two test applications, one with aspect-oriented execution tracing and the other one with conventional object-oriented approach, to trace the same amount of information. Both applications were created without the use of logging frameworks to rule out their potential performance effects in the measurements. The comparison of the 12 measurements with regard to the median of the data resulted in statistically significant (p < 0.001) differences:

1. The Java and AspectJ version influence the execution performance, Java 1.9 and, the most current, AspectJ version 1.9 performed better.
2. We have isolated the impacts of the performance cuts (a) coming from constructing the trace messages from the necessary data and (b) writing the constructed trace messages to files. Constructing the trace messages had far more impact on the performance than writing these data to a file, as shown in Section 1.3. However, this finding depends also on the tracing policy of the application as it determines the amount of data collected for the trace messages.
3. Aspect-oriented implementation of execution tracing compared to its plain object-oriented counterpart deteriorated the performance. The performance cut moved in the range of 15%–56% in our measurements while tracing was deactivated. This performance overhead depends also on the aspect-oriented trace implementation and use of join points. On the one hand, the test applications were very resource intensive in terms of CPU, probably far more than a usual application is, which magnified the performance differences; on the other hand, the aspect-oriented implementation used the this JointPoint construct to obtain function signatures and parameters. Colyer et al. in [19] managed to decrease the performance overhead with deactivated execution tracing to 1% after eliminating the use of thisJointPoint with an earlier AspectJ and Java version.

4. Aspect-oriented implementation of execution tracing caused a performance cut compared to the non-aspect-oriented implementation when tracing was activated. However, the ratio of the performance cut was smaller than with deactivated tracing as introduced in Section 1.3.

As a consequence of the measurements, the following guidelines can be articulated for a potential execution tracing implementation with regard to performance:

1. The trace messages need to be possibly short but meaningful
2. Using different trace levels is a usual way to cut the amount of data to be traced
3. Complex objects, collections should only be serialized, if they are really necessary
4. Trace messages need to be constructed, when they are necessary

None of the above recommendations are presumably new to software engineers but we would like to emphasize them as they are confirmed by our quantitative study. Nevertheless, the goal of execution tracing is to collect the necessary information for error analysis. The data has to be consistent and accurate. Boosting performance may never result in inconsistent, inaccurate, or illegible traces as it makes the use of such data then questionable. The major advantage of aspect-oriented execution tracing is that it can be added with low-cost to applications to achieve the required accuracy and consistency.

Finally, to verify and generalize our findings, the same measurements should be extended to a set of open-source applications, which possess conventionally built-in execution tracing. Then the same applications should be equipped with execution tracing in the aspect-oriented manner while the amount of information traced and the performance differences should be measured. In those cases where the amount of trace data can also significantly deviate, the quality of the produced trace data needs also be considered beside the covariates. In addition, the measurements should involve compile-time weaving as well as load-time weaving to highlight their performance effects in a quantitative manner.

References

1. Galli, T., Chiclana, F., Carter, J., Janicke, H.: Towards introducing execution tracing to software product quality frameworks. Acta Polytechnica Hungarica 11(3), 5–24 (2014). DOI 10.12700/APH.11.03.2014.03.1
2. Galli, T., Chiclana, F., Siewe, F.: Software product quality models, developments, trends and evaluation. SN Computer Science (2020). DOI 10.1007/s42979-020-00140-z
3. Uzelac, V., Milenkovic, A., Burtscher, M., Milenkovic, M.: Real-time unobtrusive program execution trace compression using branch predictor events. CASES 2010 Proceedings of the 2010 international conference on Compilers, Architectures and Synthesis for Embedded Systems, ISBN: 978-1-60558-903-9 (2010).

4. Galli, T.: Fuzzy logic based software product quality model for execution tracing. MPhil Thesis, Centre for Computational Intelligence, De Montfort University, Leicester, UK, [Online], 2013, [Accessed: 05.02.2018]. URL https://www.dora.dmu.ac.uk/bitstream/handle/2086/9736/MPhilThesisTamasGalli2013ExaminedFinal.pdf
5. Laddad, R.: AspectJ in Action. Manning, 2nd Edition (2009).
6. Breu, S., Krinke, J.: Aspect mining using event traces. In: Proceedings of the 19th IEEE International Conference Automated Software Engineering (2004).
7. Galli, T., Chiclana, F., Carter, J., Janicke, H.: Modelling execution tracing quality by type-1 fuzzy logic. Acta Polytechnica Hungarica 8(10), 49–67 (2013). DOI 10.12700/APH.10.08.2013.8.3
8. Adams, B., Schutter, K.D., Zaidman, A., Demeyer, S., Tromp, H., Meuter, W.D.: Using aspect orientation in legacy environments for reverse engineering using dynamic analysis – an industrial experience, TUD-serg-2008-035. Tech. rep., Software Engincering Research Group, Department of Software Technology, Faculty of Electrical Engineering, Mathematics and Computer Science, Delft University of Technology (2008).
9. Bispo, J., Pinto, P., Nobre, R., Carvalho, T., Cardoso, J., Diniz, P.: The MATISSE MATLAB compiler. In: 11th IEEE International Conference on Industrial Informatics (INDIN) (2013).
10. Foundation, E.: Aspectj project. Tech. rep. URL http://www.eclipse.org/aspectj/
11. SharpCrafters: Postsharp projekt. [Online], [Accessed: 30.05.2020]. URL https://www.postsharp.net/
12. Spinczyk, O., Blaschke, G., Borchert, C., Gal, A., Kramer, B., Lohmann, D., Sand, R., Schiermeier, H., Spinczyk, U., Tartler, R., Urban, M.: Aspectc++ compiler project. [Online], [Accessed: 30.05.2020]. URL http://www.aspectc.org/
13. Spinczyk, O., Lehmann, D., Urban, M.: Aspectc++: an aop extension for c++. Software Developers Journal 68–74 (2005).
14. Zaidman, A., Adams, B., Schutter, K.D., Demeyer, S., Hoffman, G., Ruyck, B.D.: Regaining lost knowledge through dynamic analysis and aspect orientation – an industrial experience report. In: Proceedings of the 10th European Conference on Software Maintenance and Reengineering, pp. 91–102 (2006).
15. Kartal, Y., Schmidt, E.: An evaluation of aspect oriented programming for embedded real-time systems. In: 22nd International Symposium on Computer and Information Sciences, (ISCIS) (2007).
16. Zhang, L.: Aspect-oriented analysis for embedded real-time systems. In: 2008 Advanced Software Engineering and Its Applications, pp. 53–56 (2008).
17. Diggins, C.: Aspect-oriented programming and c++. Dr Dobb's Journal (2004).
18. Kienzle, J., Guerraoui, R.: AOP: Does it Make Sense? – The Case of Concurrency and Failures, In Object-Oriented Programming Lecture Notes in Computer Science. Springer (2006).
19. Colyer, A., Clement, A., Bodkin, R., Hugunin, J.: Using aspectj for component integration in middleware. In: Companion of the 18th Annual ACM SIGPLAN Conference on Object-Oriented Programming, Systems, Languages, and Applications, OOPSLA '03, pp. 339–344. Association for Computing Machinery, New York, NY, USA (2003). DOI 10.1145/949344.949440
20. Davies, J., Huismans, N., Slaney, R., Whiting, S., Webster, M., Berry, R.: An aspect oriented performance analysis environment. In: Practitioners' Report from the International Conference on Aspect-Oriented Software Development (2003).

21. Mortensen, M., Ghosh, S., Bieman, J.: Aspect-oriented refactoring of legacy applications: An evaluation. IEEE Transactions on Software Engineering 38(1), 118–140 (2012).

22. Hilsdale, E., Hugunin, J.: Advice weaving in aspectj. In: Proceedings of the 3rd International Conference on Aspect-Oriented Software Development, AOSD '04, pp. 26–35. Association for Computing Machinery, New York, NY, USA (2004). DOI 10.1145/976270.976276

23. Ali, M.S., Babar, M.A., Chen, L., Stol, K.J.: A systematic review of comparative evidence of aspect-oriented programming. Information and Software Technology 52(9), 871–887 (2010). DOI https://doi.org/10.1016/j.infsof.2010.05.003

24. Kitchenham, B., Charters, S.: Guidelines for performing systematic literature reviews in software engineering. Technical Report, EBSE-2007-01 (2007).

25. Spring: Spring AOP APIs. [Online], [Accessed: 07.03.2020] (2020). URL https://docs.spring.io/spring/docs/current/spring-framework-reference/core.html#aop-api

26. Jung Pil Choi: Aspect-oriented programming with enterprise javabeans. In: Proceedings Fourth International Enterprise Distributed Objects Computing Conference. EDOC2000, pp. 252–261 (2000).

27. Šuta, E., Martoš, I., Vranic¢, V.: Usability of aspectj from the performance perspective. In: 2015 IEEE 1st International Workshop on Consumer Electronics (CE WS), pp. 83–86 (2015).

28. Salkind, N.J.: Exploring Research, 7th Edition. Sage (2008).

Appendix 1.A
A Normality Check of the Measurement Data

In this section we provide the data collected by our 12 measurements and the checks how these data approximate the theoretical normal distribution. Each measurement contains 10 test suites and each test suite contains 20 runs of the tested applications, which run in parallel to eliminate the extraneous variables as the 6 parallel pairs listed below show:

1. Java 1.8 test application with conventional tracing deactivated – Java 1.8 test application with aspect-oriented tracing deactivated
2. Java 1.8 test application with conventional tracing activated, output in/dev/null – Java 1.8 test application with aspect-oriented tracing activated, output in/dev/null
3. Java 1.8 test application with conventional tracing activated, output in real file – Java 1.8 test application with aspect-oriented tracing activated, output in real file
4. Java 1.9 test application with conventional tracing deactivated – Java 1.9 test application with aspect-oriented tracing deactivated
5. Java 1.9 test application with conventional tracing activated, output in/dev/null – Java 1.9 test application with aspect-oriented tracing activated, output in/dev/null
6. Java 1.9 test application with conventional tracing activated, output in real file – Java 1.9 test application with aspect-oriented tracing activated, output in real file

The diagrams depict the data of the test runs in each test suite on the y axis and a theoretical normal distribution with zero mean value on the x axis. If the two distributions are the same, then the points follow a straight line, which is also shown as an identity line on the below charts.
   Abbreviations used on the charts:

- Application with aspect-oriented tracing
- Application with conventional tracing
- Execution tracing deactivated
- Execution tracing activated

1.A.1 Java 1.8, Test Application with Conventional Tracing Switched Off

Table 1.A.1 Runtime of the Java 1.8 test application in msec., with conventional tracing, trace switched off

ID	Suite 1	Suite 2	Suite 3	Suite 4	Suite 5	Suite 6	Suite 7	Suite 8	Suite 9	Suite 10
1	7922	7927	7982	7949	8015	7953	7952	7910	7915	7951
2	7832	7772	7801	7823	8272	8061	7848	7960	7917	7850
3	7995	7919	7946	7718	7890	8075	7843	7807	7855	8311
4	8005	8153	8168	7842	8077	8244	7796	8238	7866	8201
5	7941	7985	8042	7927	7984	8947	7960	7899	7889	8086
6	7887	7894	7978	7919	8135	8173	7915	7812	7754	7920
7	7960	7794	7830	7854	8097	8134	8159	8093	7906	8005
8	8023	7973	7945	7895	7961	8061	7905	8042	8068	7880
9	7947	8100	12570	8006	7659	8111	7826	7990	7961	7923
10	7919	7935	8146	8193	7766	7890	7823	7808	8530	7740
11	7870	7877	8526	8040	7817	8068	8240	7816	8072	7809
12	7764	7837	11100	7938	7748	7971	7887	7879	7797	7811
13	7851	7717	8913	7818	7849	7962	7985	7838	8203	7659
14	7910	7985	11280	8691	7840	7833	9286	7896	7950	7790
15	8022	8121	8596	8126	7909	7862	8024	8183	8353	7887
16	7950	7818	9887	8004	8008	7913	7981	7949	8092	7876
17	7947	7805	9323	7787	7900	8037	7893	7848	7907	7806
18	7702	7897	10270	7836	7825	8062	7790	8251	8163	7761
19	8113	8220	8058	7939	8052	7828	8052	7862	7907	7915
20	7909	7960	7846	8045	7910	7944	7881	7947	7995	7833

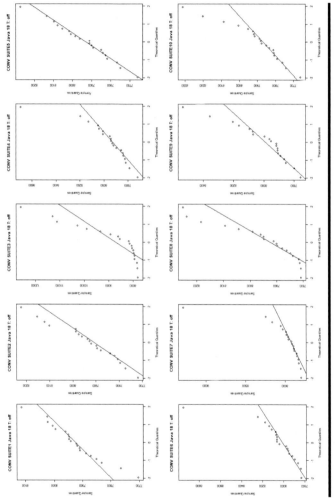

Figure 1.A.1 Quantile–quantile plot of the runtimes for each test suite, java 1.8, conventional tracing, trace switched off.

1.A.2 Java 1.8, Test Application with Conventional Tracing Switched On, Output in/dev/null

Table 1.A.2   Runtime of the Java 1.8 test application in msec., with conventional tracing, trace switched on, output in /dev/null

ID	Suite 1	Suite 2	Suite 3	Suite 4	Suite 5	Suite 6	Suite 7	Suite 8	Suite 9	Suite 10
1	48886	47283	44654	49118	49829	47635	44882	50651	45976	48138
2	48231	45584	47406	49642	53092	45899	48312	49945	50222	48741
3	44431	46552	45036	47521	45530	46550	48077	46942	48036	47498
4	47027	45681	47901	46221	44424	43727	48092	45361	47397	45992
5	45806	43758	50973	45704	46438	46954	49493	44398	45799	43319
6	47719	46700	47302	49661	46154	47370	49380	44191	47147	49990
7	48990	48284	45530	44122	45443	48673	45305	49720	49065	45638
8	48044	50027	43872	47387	49462	45623	48708	46217	49622	43874
9	49193	48984	46091	47117	44993	45665	47545	50466	48278	45187
10	45735	44494	44886	48157	42944	47264	42631	44377	49234	47751
11	46433	46626	51061	44549	46837	45545	46784	46473	44959	48182
12	46337	48067	51139	46601	45596	46274	48122	47192	44276	42625
13	47107	45553	47829	48671	44135	45948	45911	42262	45903	44629
14	47319	48566	45503	46379	46165	50941	47157	45915	43025	47827
15	46326	45317	44723	45808	46577	47037	45901	45799	48784	46775
16	44819	49660	47086	42463	48938	48149	47524	46463	45768	44565
17	46840	48660	45606	44921	46024	50124	44993	49032	52209	46034
18	49478	46270	51535	48108	43208	47417	44888	47986	47277	45809
19	47380	46537	48395	49393	46755	44855	47791	43526	47663	46404
20	49562	46408	47017	46802	43500	46596	63860	45573	49725	44520

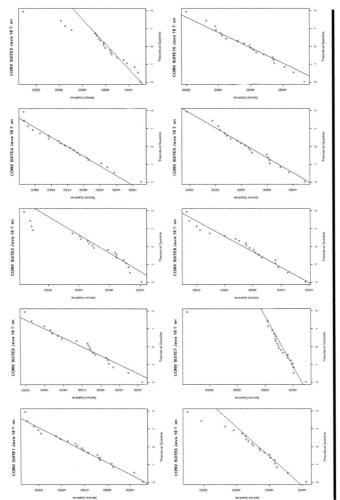

Figure 1.A.2 Quantile–quantile plot of the runtimes for each test suite, Java 1.8, conventional tracing, trace switched on, output in /dev/null.

1.A.3 Java 1.8, Test Application with Conventional Tracing Switched On, Output in Real File

Table 1.A.3   Runtime of the Java 1.8 test application in msec., with conventional tracing, trace switched on, output in real file

ID	Suite 1	Suite 2	Suite 3	Suite 4	Suite 5	Suite 6	Suite 7	Suite 8	Suite 9	Suite 10
1	62525	64618	67214	70127	61690	68995	69197	67602	68876	68004
2	68783	64393	68635	63133	64762	70993	68080	73609	69449	64392
3	66545	70246	66551	64019	65725	68433	66818	63320	64611	65582
4	67874	61270	63776	68398	64441	63720	65497	63361	66338	68070
5	62769	62447	62067	64546	64408	63060	64779	65199	66153	62243
6	63962	67142	66511	68590	66795	60958	61237	64088	64029	66652
7	65734	64747	64798	62923	66201	67241	66832	71164	65436	67232
8	68803	63471	66641	66337	64260	66720	67222	65119	65495	62774
9	65783	63910	67391	63239	66976	65992	64287	69496	66707	69460
10	67246	66888	66109	61077	63539	62145	64744	70081	63845	65157
11	64237	69633	61718	65367	63371	69553	63162	68685	66778	64090
12	63657	67660	64843	61701	60948	61985	64961	68104	65944	65031
13	65247	63043	63885	65808	59993	61127	65465	66416	64680	60640
14	67202	64887	62581	61993	63315	65203	61850	65098	67322	65882
15	63579	72244	65390	63070	67395	70346	68619	63406	64369	67425
16	63384	63737	67576	65690	60151	102572	65046	71923	62137	64190
17	66093	67760	64371	62845	64467	64717	65421	64877	64378	64646
18	65228	67469	64306	67650	67580	64831	62584	63732	69372	63042
19	65577	66065	62429	68234	65758	62782	62765	63542	67118	70645
20	63320	65253	64649	69128	63366	70228	59868	66242	66581	63632

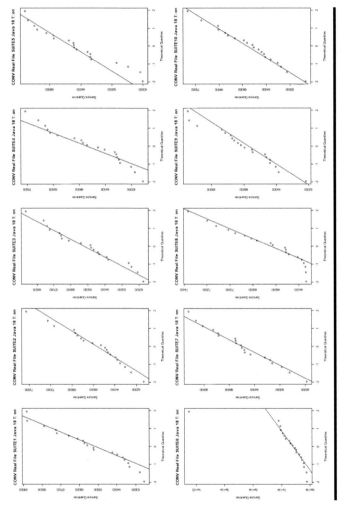

Figure 1.A.3 Quantile–quantile plot of the runtimes for each test suite, Java 1.8, conventional tracing, trace switched on, output in real file.

1.A.4 Java 1.9, Test Application with Conventional Tracing Switched Off

Table 1.A.4 Runtime of the Java 1.9 test application in msec., with conventional tracing, trace switched off

ID	Suite 1	Suite 2	Suite 3	Suite 4	Suite 5	Suite 6	Suite 7	Suite 8	Suite 9	Suite 10
1	4613	4731	4720	4737	4665	4686	4659	4677	4736	4630
2	4528	5570	4627	4679	4613	4632	5290	4684	4507	4741
3	4807	4763	4801	4654	4644	4686	4617	4795	4784	4617
4	4593	4492	4693	4483	4476	4458	4612	4548	4579	4504
5	4683	4639	4593	4749	4603	4565	4643	4899	4589	4557
6	4574	4535	4608	4614	4689	4478	4533	4491	4492	4591
7	4631	4621	4995	4755	4829	4920	4670	4457	4664	4629
8	4533	4551	4887	4550	4535	4549	4488	4627	4518	4765
9	4603	4663	5049	4663	4444	4968	4597	4879	4865	4612
10	4513	4597	4749	4594	4489	4496	4615	5041	4853	4667
11	4673	4666	4638	4590	4587	4702	4598	4853	4630	4662
12	4439	4570	4599	4658	4660	4600	4609	4681	4713	4592
13	4505	4672	4618	4750	4698	4688	4775	4556	4527	4890
14	4560	4557	4619	4570	4586	4740	4727	4536	4588	4741
15	4581	4685	4633	5214	4585	4605	4537	4828	4697	4645
16	4808	4614	4581	4751	4784	4702	4656	4559	5097	5107
17	4723	4580	4747	5131	4614	4764	4631	4588	4750	4898
18	4656	4609	4580	4649	4631	4519	5001	4426	4584	4607
19	4678	4652	4686	4598	4697	4616	4810	4684	4710	4868
20	4606	4824	4604	4692	4523	4556	4692	4568	4600	4648

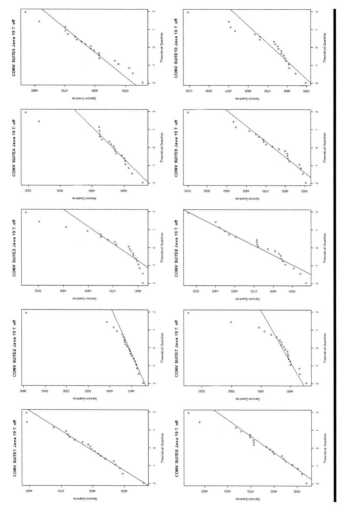

Figure 1.A.4   Quantile–quantile plot of the runtimes for each test suite, Java 1.9, conventional tracing, trace switched off.

1.A.5 Java 1.9, Test Application with Conventional Tracing Switched On, Output in/dev/null

Table 1.A.5   Runtime of the Java 1.9 test application in msec., with conventional tracing, trace switched on, output in /dev/null

ID	Suite 1	Suite 2	Suite 3	Suite 4	Suite 5	Suite 6	Suite 7	Suite 8	Suite 9	Suite 10
1	32232	32503	31119	34561	31357	32579	34055	32334	32720	31788
2	33698	32209	34469	33530	33008	30820	32317	30828	30667	31958
3	31285	32545	34294	32047	31234	32674	32370	30840	31357	31907
4	31999	30308	32781	32722	33955	32933	32044	34740	31541	36248
5	29731	31037	30317	31142	32207	30953	30574	33647	33819	32260
6	33425	30663	31404	32084	29929	31394	31503	31247	39138	32749
7	32685	34250	32437	31887	33264	31108	32121	32482	35718	31990
8	32909	35121	32370	32723	31809	30934	31078	32367	36739	30891
9	31776	35122	33438	33479	31792	33701	34463	33658	34060	32533
10	31907	31143	30921	30608	31410	33354	31868	32242	32795	32347
11	34905	30565	32217	30369	31724	31445	29797	29677	33180	32954
12	31077	32234	32850	34373	32690	30902	33084	31415	33149	29903
13	34443	32132	30605	31389	32974	32228	33101	31776	35271	33479
14	33425	32270	32400	32978	32662	31767	32425	33028	33963	31177
15	32507	31545	30843	31846	33307	33507	32220	32620	31846	32628
16	29699	30622	31028	31016	31855	31563	29833	32349	32039	29818
17	31440	30987	33483	32030	31066	31189	31511	30858	32629	31541
18	32827	33427	33439	31690	32972	31561	33137	32509	30527	31223
19	32612	31735	33186	32106	32079	33319	33849	32193	34013	34518
20	29669	32325	32109	34185	32020	32523	32319	32053	36530	33852

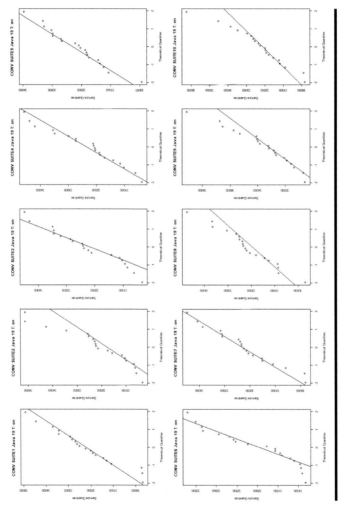

Figure 1.A.5   Quantile–quantile plot of the runtimes for each test suite, Java 1.9, conventional tracing, trace switched on, output in /dev/null.

1.A.6 Java 1.9, Test Application with Conventional Tracing Switched On, Output in Real File

Table 1.A.6 Runtime of the Java 1.9 test application in msec., with conventional tracing, trace switched on, output in real file

ID	Suite 1	Suite 2	Suite 3	Suite 4	Suite 5	Suite 6	Suite 7	Suite 8	Suite 9	Suite 10
1	50643	51692	52946	54366	52445	50302	51414	49650	50487	53153
2	51991	48443	50723	49339	48197	48116	49189	51961	52705	48423
3	51460	51256	50884	49398	47706	54491	48268	48992	48732	48866
4	51093	50756	48183	52449	51594	51832	50340	47374	54681	48629
5	48841	50318	49156	48280	51339	48283	50388	50436	51557	50818
6	48886	51625	47810	52291	50625	49491	49560	48762	50108	52147
7	49806	50865	51067	49886	48364	49780	51335	49000	50677	46248
8	48950	48250	48263	47034	49151	48311	50016	50015	48046	48982
9	49028	48868	50220	50803	50161	50282	51238	49904	49333	49606
10	50980	51735	49804	48469	50467	47099	51631	52397	48070	48774
11	51499	54608	51606	51405	51155	49789	48894	50146	51830	50235
12	49126	48740	48061	50412	50569	49961	51071	50599	47575	47088
13	48460	50265	48658	50510	49205	53135	52552	50628	50216	54358
14	48872	56077	49802	51685	49761	53407	48649	48648	49246	50048
15	47325	48867	51368	48217	50712	51108	49525	45717	47648	47778
16	50915	50742	49378	52298	52572	49916	47969	48284	51482	47923
17	51331	53833	53845	48602	49373	48925	50825	51278	51050	51753
18	53312	50026	49547	51214	49932	49415	49798	49773	51288	48830
19	47510	49833	52072	49269	50309	51198	49945	49387	49051	50972
20	48325	49588	49304	51125	47647	51120	48190	51695	51235	48445

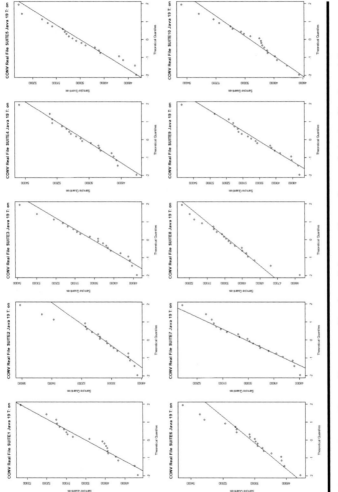

Figure 1.A.6 Quantile–quantile plot of the runtimes for each test suite, Java 1.9, conventional tracing, trace switched on, output in real file.

1.A.7 Java 1.8, Test Application with Aspect-Oriented Tracing Switched Off

Table 1.A.7   Runtime of the Java 1.8 test application in msec., with AO-tracing, trace switched off

ID	Suite 1	Suite 2	Suite 3	Suite 4	Suite 5	Suite 6	Suite 7	Suite 8	Suite 9	Suite 10
1	9311	9257	9209	9056	9739	9398	9082	9718	9110	9364
2	9069	9269	9332	9118	9695	9284	8938	9064	9121	9648
3	9193	9320	9154	9015	9300	9584	9137	9110	9220	9793
4	9265	8887	9073	9086	9106	9103	9080	9337	9023	9312
5	9133	9313	9308	9363	9163	9962	9223	9076	9161	9352
6	8940	9171	8996	9301	9496	9323	9220	9241	9108	9225
7	9086	9134	9105	9040	9978	8876	9212	8850	9129	9229
8	9154	9285	12346	9134	9047	9243	9251	9118	9847	9167
9	9428	9124	11067	9533	9172	9120	9279	9051	9447	9338
10	9206	9061	11117	9068	9156	9285	9291	9311	9031	9460
11	9247	9236	12343	9184	9257	9084	9043	9204	9312	9194
12	9377	10535	13319	9622	9332	9366	9048	9353	9003	9259
13	8912	9296	13512	9811	9358	9275	9289	9088	9106	9286
14	8922	9298	11355	9191	9178	9092	9087	9434	8891	9191
15	9212	9300	13122	9223	9144	9324	9192	9453	8949	9174
16	9262	9361	9181	9205	9129	9159	9218	9292	8934	9245
17	9104	8998	9529	9229	9318	9208	9170	9074	9169	8986
18	10049	10238	10920	9930	9775	9952	9927	10152	9575	10269
19	10408	10488	10138	9820	10449	10707	9994	10141	9909	10359
20	10195	10214	10561	9974	9502	10066	9742	9950	9352	9503

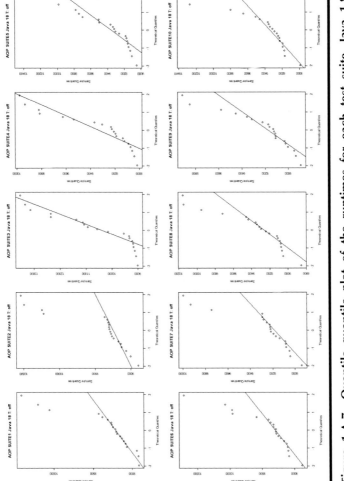

Figure 1.A.7 Quantile–quantile plot of the runtimes for each test suite, Java 1.8, AO-tracing, trace switched off.

1.A.8 Java 1.8, Test Application with Aspect-Oriented Tracing Switched On, Output in/dev/null

Table 1.A.8 Runtime of the Java 1.8 test application in msec., with AO-tracing, trace switched on, output in /dev/null

ID	Suite 1	Suite 2	Suite 3	Suite 4	Suite 5	Suite 6	Suite 7	Suite 8	Suite 9	Suite 10
1	50893	47486	45563	58389	55731	51265	53158	54038	54817	52922
2	51805	50212	52308	48387	54508	49349	49205	58363	52038	60261
3	47915	48330	46160	51774	47247	43527	54516	47045	49363	46125
4	52961	49350	50805	47597	46470	43892	49646	45568	44712	48995
5	52968	47444	50498	47931	49015	44225	51506	45205	54929	45281
6	46369	49867	47521	51636	50165	49652	54009	52482	48667	53588
7	55020	54199	56132	48082	48632	49804	47683	51617	53063	45771
8	53639	54100	45977	47137	50775	50451	57050	49905	54572	50819
9	48857	50364	46409	48726	50700	48112	52808	53629	46398	44917
10	46343	50286	48555	52555	44944	47184	48943	50151	50462	54372
11	47975	47368	56027	48233	53542	46068	54124	45365	52206	51074
12	50800	52890	56389	49851	49983	48031	49898	49185	47448	49336
13	49066	53458	48950	51865	57902	47736	50014	48507	49176	57873
14	47211	50034	48242	48270	47113	54485	48566	50588	53357	45000
15	50151	48975	50833	48537	54702	52166	48964	47793	51009	45550
16	47180	50330	50097	50482	46448	53350	47585	49619	55200	47168
17	53310	49327	55562	53219	44018	52860	48968	49221	47698	50197
18	47080	51026	52772	52881	50885	50354	61221	48389	51672	47090
19	50863	45446	46685	46954	48537	46025	61479	48440	54804	51896
20	51123	55064	49723	52034	54142	56130	54753	53564	54599	49853

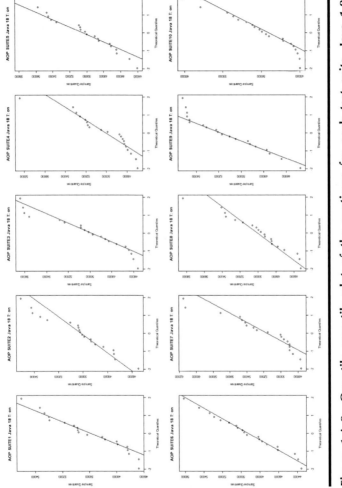

Figure 1.A.8 Quantile–quantile plot of the runtimes for each test suite, Java 1.8, AO-tracing, trace switched on, output in /dev/null.

1.A.9 Java 1.8, Test Application with Aspect-Oriented Tracing Switched On, Output in Real File

Table 1.A.9 Runtime of the Java 1.8 test application in msec., with AO-tracing, trace switched on, output in real file

ID	Suite 1	Suite 2	Suite 3	Suite 4	Suite 5	Suite 6	Suite 7	Suite 8	Suite 9	Suite 10
1	62846	70937	72871	71141	70480	76028	71843	70875	75226	76878
2	71985	70876	77594	68962	70195	73689	68957	77126	71673	68674
3	62627	69321	67901	72014	69128	72512	70244	69503	72145	66073
4	75417	63327	66562	69539	70987	70462	66553	68441	70860	68517
5	64416	64142	69259	63857	70565	63498	65194	69000	70576	71799
6	68998	72236	75603	71669	66077	62586	69062	69945	70851	72077
7	72137	70025	67914	64213	71958	71446	66537	74209	64351	71452
8	75698	67889	71987	65101	77387	72301	74418	70150	64823	69782
9	67842	68431	71200	70001	69111	71212	69712	74465	73259	75853
10	72951	69221	64530	70153	65149	67580	70461	75327	67990	74458
11	67636	68676	62393	70752	68102	72527	69405	68443	72654	64010
12	68695	68088	68285	67969	64525	66220	70564	71986	65614	69624
13	69339	64134	66399	70836	67909	71993	70158	64640	67167	72101
14	70445	72143	69298	68180	71387	73711	64077	68104	71460	73096
15	65436	70592	77637	71009	66949	107072	69847	70408	65807	73485
16	69295	65420	72234	65111	70724	65389	68439	69716	69443	73407
17	67674	72728	68335	66353	65846	69700	70078	68577	75561	66244
18	71800	68229	67113	71588	68552	68943	65941	68297	71815	73780
19	66398	70439	70212	76179	70950	74297	61539	73900	66706	68795
20	65627	66970	64037	63949	55548	62831	63225	62468	65239	67303

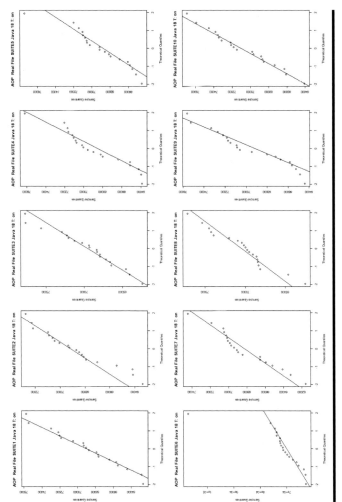

Figure 1.A.9 Quantile–quantile plot of the runtimes for each test suite, Java 1.8, AO-tracing, trace switched on, output in real file.

1.A.10 Java 1.9, Test Application with Aspect-Oriented Tracing Switched Off

Table 1.A.10   Runtime of the Java 1.9 test application in msec., with AO-tracing, trace switched off

ID	Suite 1	Suite 2	Suite 3	Suite 4	Suite 5	Suite 6	Suite 7	Suite 8	Suite 9	Suite 10
1	7487	7153	7279	7187	7294	7284	8597	7157	7065	7200
2	7280	7733	7061	6902	7218	7780	7447	7362	7062	7258
3	7107	7213	7356	6950	7142	7440	7083	7145	7505	7087
4	6998	6999	7008	6868	7206	7043	7113	7008	7101	7021
5	7346	7130	7140	6832	7104	7128	7038	7273	6965	7357
6	6881	7223	7400	7420	7175	7052	7232	7262	7087	7493
7	7007	6913	7906	6716	6959	7079	7054	7526	6905	7125
8	7278	7080	6907	7018	7190	6836	6827	7485	6976	7283
9	7072	6995	7374	7000	6988	7528	7138	7152	7071	7009
10	7201	7016	7047	7110	7384	6770	7057	7256	7099	7588
11	6873	6920	7034	7458	7601	6920	7217	7070	7058	7271
12	7238	7162	7107	6872	6982	7085	7061	7188	7522	7738
13	6974	7430	7221	7496	7013	7621	7098	7279	7174	7323
14	7284	7501	7753	7522	7309	7342	7834	7299	7389	7617
15	7002	7284	7307	7334	7364	7170	7382	6748	7284	7049
16	7349	7092	7283	7356	7213	7063	7086	7087	7325	7053
17	7049	7533	7053	7610	7186	7277	7314	6771	7338	6886
18	7230	7096	7203	7285	7288	7214	7275	7006	8098	6604
19	7176	7090	7297	7654	7145	7469	7208	7144	7455	7121
20	7192	7245	7294	7704	7724	7685	7191	6966	7563	7046

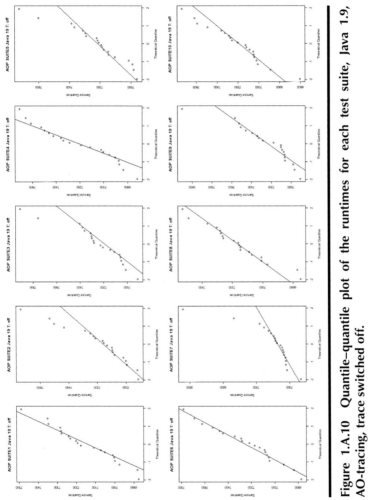

Figure 1.A.10 Quantile–quantile plot of the runtimes for each test suite, Java 1.9, AO-tracing, trace switched off.

1.A.11 Java 1.9, Test Application with Aspect-Oriented Tracing Switched On, Output in/dev/null

Table 1.A.11 Runtime of the Java 1.9 test application in msec., with AO-tracing, trace switched on, output in /dev/null

ID	Suite 1	Suite 2	Suite 3	Suite 4	Suite 5	Suite 6	Suite 7	Suite 8	Suite 9	Suite 10
1	40055	39050	41731	41386	38905	38719	40868	40388	40615	38641
2	41079	38333	44264	44167	37644	38677	38376	38472	37635	37619
3	38939	43356	38628	41925	39249	39055	43726	39072	38618	38273
4	38983	37039	38048	40658	39830	39383	37334	38027	42193	44570
5	39673	41566	40790	39130	40448	39509	39668	40120	46607	37254
6	37937	40410	39811	38048	37450	37473	38085	39396	42843	42534
7	37519	43438	38876	39322	39880	38055	38822	38571	42718	39212
8	36795	40253	40994	38012	38871	39627	38920	44157	39206	37276
9	35942	39192	38593	40977	37044	38190	40479	37514	39728	40137
10	39517	38887	38504	38950	40917	42426	43142	37624	45592	38778
11	43305	36858	38804	37506	38972	36728	39388	40574	40696	40711
12	39085	35381	38989	40918	40789	38834	37416	38728	39428	37902
13	40695	41889	37934	39575	38320	42422	38691	37503	38473	39645
14	41093	40110	40383	38253	39033	40489	38935	42065	39042	40381
15	38615	38342	40770	39850	40144	40772	40031	38560	39294	39772
16	39999	37989	43537	41155	38548	40985	38611	40089	39202	43154
17	37597	39201	37856	37938	37277	39442	39190	38484	46377	38164
18	37177	39049	36641	36255	35570	35992	38167	37096	40329	36620
19	37697	37111	36766	36226	35694	37361	37661	38889	36989	37347
20	38368	37968	37845	37699	39731	37211	36457	39274	37088	36871

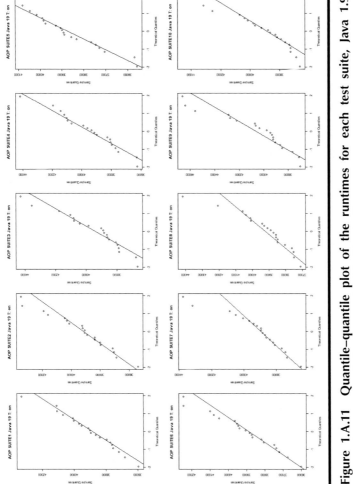

Figure 1.A.11 Quantile–quantile plot of the runtimes for each test suite, Java 1.9, AO-tracing, trace switched on, output in /dev/null.

1.A.12 Java 1.9, Test Application with Aspect-Oriented Tracing Switched On, Output in Real File

Table 1.A.12   Runtime of the Java 1.9 test application in msec., with AO-tracing, trace switched on, output in real file

ID	Suite 1	Suite 2	Suite 3	Suite 4	Suite 5	Suite 6	Suite 7	Suite 8	Suite 9	Suite 10
1	61669	60546	58933	59484	60875	60543	60890	59785	61599	59510
2	56796	56582	55187	56527	54877	60812	56441	59524	56695	58710
3	57702	55603	54169	57865	58460	56437	56532	59235	56756	57941
4	58186	59380	58491	57952	61579	54901	56445	56629	63685	60381
5	57858	56324	55532	57261	58610	55396	58491	55694	56070	58456
6	56784	58184	56655	58200	60780	60244	63387	58453	57009	56349
7	57157	56423	60515	56420	61044	56444	58257	58098	55882	55128
8	55886	60421	56412	56493	57901	56796	58165	58414	58300	57613
9	62220	56010	56925	56602	60045	56490	58865	55039	55282	55676
10	59168	59725	61258	56725	57048	58785	55562	56578	59225	57377
11	56845	55083	61504	61139	59424	58424	60067	58789	55274	55710
12	59575	55639	61824	59100	58410	57603	56502	58129	59158	55707
13	55836	58421	57523	60360	59150	58034	57737	58106	58253	56193
14	60967	58423	60325	62091	58803	62643	62320	58566	62184	59432
15	57212	58675	59493	56202	57717	57192	59117	58140	59664	57718
16	57797	60272	61055	58972	63929	57490	56496	58163	57268	56702
17	57915	57327	61229	57245	60437	59954	57332	55242	57387	57290
18	53694	58555	53105	56366	51863	57608	53081	57382	55160	56116
19	52516	51618	53748	53116	51258	52391	51734	54444	55314	53662
20	54794	52984	53689	54719	52429	56236	52762	54709	55619	55396

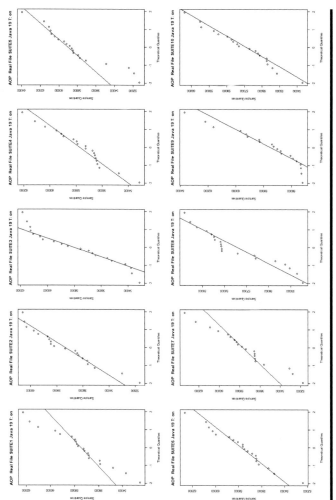

Figure 1.A.12 Quantile–quantile plot of the runtimes for each test suite, Java 1.9, AO-tracing, trace switched on, output in real file.

Chapter 2

A Survey on Software Test Specification Qualities for Legacy Software Systems

Shilpa Mariam George[a],
Saradha Yasasvi Yeduruvada[a],
and Dae-Kyoo Kim[a]

[a]Oakland University, Rochester, Michigan

Contents

2.1	Introduction	60
2.2	Related Work	61
2.3	Characteristics in Legacy Systems	62
2.4	Qualities of Test Specifications	63
2.5	Survey	65
2.6	Threats to Validity	71
	2.6.1 Population Validity	72
	2.6.2 Content Validity	72
	2.6.3 Data Extraction Validity	72
2.7	Conclusion	72
References		73

2.1 Introduction

A legacy system is a software solution whose languages, codes, standards, and technologies belong to a prior era of innovation [1]. Legacy systems are still widely used across various sectors including health care, banking, enterprise, energy, military, and government. Legacy systems are usually critical to the day-by-day operations of the organization, but difficult to maintain due to incomplete or outdated documentation about the system and the difficulty in finding needed expertise. In many cases, the maintenance and operation of a legacy system rely on the succeeded knowledge of users, which is error-prone and thus, increases maintenance cost and accelerates the deterioration of the system.

Legacy systems may evolve in three ways [2] – maintenance, modernization, and replacement. Maintenance allows the sustention of the business, but the system becomes increasingly deteriorated over time. When it reaches the point where maintenance eventually falls behind to satisfy the business needs, the system needs to be modernized. Modernization involves migrating the system to modern technologies, which is a complex and time-consuming process. It also involves a high risk in business operations as legacy systems are usually vital to the enterprise. When the system can no longer evolve, it must be replaced. As the system goes through the evolution process, significant testing is involved in every point of evolution activities.

For quality testing, it is essential to have quality test specifications. Developing quality test specifications requires a clear understanding of the quality aspects that are concerned in test specifications. Despite the importance of test specification qualities, there is little study on the quality aspects of test specifications for legacy systems. In this chapter, we present a survey on qualities of test specifications that are desired for legacy systems. The survey was prepared based on the guidelines by Punter et al. [18]. The survey includes a set of test specification qualities for rating with open fields where participants can provide comments on other qualities that are considered as important in the field. Also, the survey asks for comments on the factors that affect the qualities of test specifications and major difficulties in developing quality test specifications. Software engineers who have many years of experience in legacy systems from diverse domains are invited to the survey. The organizations from which the participants are invited include IBM, Cognizant, Tata Consultancy Services, and others. We analyzed the responses and identified concerned qualities for test specifications of legacy systems in the order of significance. We also discuss other qualities that practitioners consider as important from their experiences and influencing factors and major difficulties in developing quality test specifications. The main contributions of this work are (i) identification of the most concerned qualities of test specifications for legacy systems, (ii) identification of factors affecting the creation of good test specifications, and (iii) identification of challenges hindering the development of good test specifications from an industry perspective.

The rest of this chapter is organized as follows. Section 2.2 presents the related work. Section 2.3 discusses the characteristics of legacy systems. Section 2.4 describes the qualities concerned in the development of test specifications. Section 2.5 describes the process of the survey and the results. Section 2.6 concludes the chapter.

2.2 Related Work

Comella-Dorda et al. [3] presented a survey study on modernization techniques for legacy systems. The techniques surveyed include screen scraping, database gateway, XML integration, database replication, CGI integration, object-oriented wrapping, and componentization of legacy systems. These techniques are analyzed in terms of source artifacts to be modernized, target artifacts, strengths, and weaknesses. Lee et al. [4] presented a survey study on software testing practices and opportunities in terms of software testing methods and tools (STMT) at different levels (i.e., low, limited, lack) of use. Their study also discusses limited interoperability between methods and tools and the need for description of STMT.

The work by Jovanovikj et al. [5] presented an approach for evaluating the quality of test cases in the context of their use. Built upon the Goal-Question-Metric (GQM) approach, their work provides a process for establishing a quality plan for the quality evaluation of test cases based on the characterization of the context in which test cases are measured. The context factors used for characterization include the environment, the domain, and the associated artifacts. Their work can be beneficial to better understand legacy systems in terms of context-specific qualities, which are usually not considered in practice.

The work by Adlemo et al. [8] presented 15 qualities that constitute "good" test cases. They defined these qualities and prioritized them by order of importance based on a survey among Swedish software experts. Their work serves as a reference for test case qualities in general. However, they do not consider the specific nature or characteristics of any software systems, which in turn has an impact on the prioritization of its test case qualities. In our work, we take into account the characteristics of legacy software systems and aim to specifically address the quality concerns pertaining to such systems.

The work by Hussain et al. [6] discusses what a legacy system is and what are its components. They discuss the various problems and concerns involved in maintaining legacy applications. They also come up with ways to revive legacy systems such as software reuse, building a new system, reengineering, reverse engineering, restructuring, and forward engineering. But they do not consider the importance of having good test cases while enhancing and preserving a legacy application. They also overlook the limited knowledge about a legacy application while pointing out major challenges in the maintenance of legacy systems.

The work by Nirmala and LathaMaheswari [7] presents an approach for automatically generating test cases. They also discuss how test cases can be developed to improve the consistency, correctness, and reusability of existing software. Their work helps identify relevant test case qualities and their importance. They also describe certain factors that help in the creation of good test cases. The chapter states the understanding of requirements as a key factor that aids in the development of quality test cases.

The work by Adlemo et al. [8] presented 15 qualities of test specifications, which are prioritized by importance based on a survey among Swedish software experts. While their work serves as a reference for test specification qualities in general, it does not address quality concerns specific to legacy systems. The work by Hussain et al. [6] discusses different types of components involved in legacy systems and related concerns. However, the concern on limited knowledge of legacy systems is not addressed, which is a major challenge in the maintenance of a legacy system. The work by Nirmala and LathaMaheswari [7] presents an approach for automatically generating test specifications to improve the consistency, correctness, and reusability of existing software.

2.3 Characteristics in Legacy Systems

Legacy systems are described as outdated computer systems that encapsulate the existing business processes, organization structure, culture, and information technology [9]. They are vital to the organization but difficult to cope with [10]. Typical examples of legacy systems are those that are written in old languages such as FORTRAN and COBOL running on obsolescent machines (e.g., mainframes). In order to develop quality test specifications, the characteristics of legacy systems should be clearly understood. The following describes the main characteristics of legacy systems based on the study of existing work [11].

- ■ *Mission critical:* Legacy systems are usually the backbone of an organization's business and the main vehicle for business operations. Therefore, their failure may cause a serious impact on business.
- ■ *Brittle and inflexible to changes:* In the era where legacy systems were developed several decades ago, there rarely existed any rigorous development processes, established methods, or solid development tool support. So, the development of legacy systems was very lenient and *ad hoc*, which makes them brittle and inflexible to changes.
- ■ *Running on obsolete hardware:* Legacy systems are usually running on outdated hardware that is slow and expensive to maintain. When developed, they are designed specifically for the hardware and it is very difficult (if not impossible) to run them on new hardware.
- ■ *Incomplete or outdated documentation:* Due to the lack of rigorous development processes and established methods when legacy systems were developed, there exists little documentation about their development. The

absence of documentation becomes a major challenge in the maintenance of legacy systems.

■ *Difficult to find expertise:* As legacy systems were developed decades ago, the developers who participated in their development or have internal knowledge (e.g., designs, behaviors) about the systems are extremely difficult to find.

■ *Expensive maintenance:* Legacy systems are expensive due to the lack of documentation and understanding of system details. This further leads to increased cost for maintenance (e.g., adding functionalities, debugging errors).

■ *Lack of clean interfaces:* Legacy systems usually do not have clean interfaces defined with other systems, which makes them difficult to extend and integrate with other systems.

■ *Outdated technologies:* The technologies used in legacy systems are outdated. There are not many technical experts available who can work on such technologies at present. It is even more difficult to find those who have enough knowledge in the specific application domain.

2.4 Qualities of Test Specifications

A test specification is a documentation of inputs, expected output, and a set of conditions under which the test item is executed [12]. To produce quality test specifications, the quality aspects of test specifications should be clearly understood. The following describes qualities concerned for test specifications.

■ *Correctness:* This quality denotes the correctness of test specifications with respect to the requirements of the concerned system and test purposes [13]. A test specification is considered as correct only when the system returns consistently correct output in a reachable end state.

■ *Clarity:* Test specifications should be clear to understand. This can be achieved by good documentation on environmental setting, input, output, and step-wise execution. This attribute supports test automation and is also known as compressibility and understandability.

■ *Repeatability:* This quality is concerned with reproducing consistent test results [8]. If test results are inconsistent, it is difficult to locate defects in the system. Test repeatability requires high clarity of test specifications.

■ *Reusability:* Test specifications should be reusable in testing similar functionalities [13]. In practice, test specification or their parts (e.g., test case, test data) are often reused. For example, test cases used for integration testing may be reused for regression testing. This quality can be achieved only when the design of test specifications is based on the set of similar functionalities to be tested. Thus, this quality requires analyzing and identifying similar functionalities before test specifications are developed.

■ *Specificity:* A test specification should be specific to the requirement that is aimed to test. This quality is particularly important for test specifications of legacy systems as code-level details are usually not available. For non-legacy systems, test specifications focus on code-level coverage (e.g., statement coverage, branch coverage). Code coverage is possible due to the availability of internal knowledge of the system. However, the same is not true for legacy systems due to the lack of documentation. Thus, a test should focus more on high-level requirements. This is particularly important when a new functionality is added, which requires testing the existing functionalities against the corresponding requirements that are impacted by the change.

■ *Accuracy:* Test specifications should be accurate in terms of test conditions, input, and expected output. This should not be confused with clarity. A test specification may be clear, but not necessarily accurate. However, clarity helps improve accuracy. The absence of accuracy in test specifications leads to inefficient testing and increasing test cost.

■ *Maintainability:* Test specifications should be stable, modifiable, and analyzable [13]. Stable test specifications allow consistent execution of tests, while modifiable test specifications give flexibility to accommodate environmental or requirement changes. Test specifications should be analyzable to verify qualities (e.g., consistency, correctness) as they evolve.

■ *Coverability:* Test specifications must have complete coverage on code changes. Types of code coverage include statement coverage, edge coverage, and condition coverage. Depending on the type of the target coverage, test specifications should be designed differently.

■ *Traceability:* Test specifications should be traceable to corresponding requirements. A common practice to achieve this quality is to explicitly specify the target requirement in the test specification. This quality is different from specificity in that while specificity is concerned with the contents of test specifications (e.g., test data, inputs, outputs) against the target requirement, traceability focuses on the mapping between test specifications and requirements.

■ *Completeness:* The test suit must consist of test specifications that together fully cover the given requirements. This quality is not for individual test specifications, but for test suits as a collection of test specifications. Therefore, it should not be confused with coverability, which is for individual test specifications covering specific requirements. High coverability does not necessary guarantees high completeness.

■ *Consistency:* Test specifications for similar functionalities should be written consistently [8], which helps reduce errors in test specifications and promote the reusability of test specifications.

■ *Efficiency:* Test specifications should be able to identify errors efficiently in the target functionality [13]. Efficiency comes in two ways – finding more errors with a smaller number of test specifications and executing test specifications with less time and resources.

■ *Autonomy:* Test specifications should have minimal dependency on each other for better efficiency. An example of a dependency is the output of a test specification being used in the input of another test specification, which imposes a constraint on the order of execution. Given that, this quality enables the execution of test specifications in a flexible order with minimal impact on the overall evaluation of the test suite [14].

■ *Conciseness:* Test specifications should be as concise as possible to improve clarity and reduce complications. This also helps improve the maintainability of test specifications.

2.5 Survey

A survey was conducted to identify the qualities that are important to legacy systems among a group of software engineers who have worked on legacy systems. The survey was prepared based on the guidelines by Punter et al. [18]. As per the guidelines, initially the survey goals were determined and then questions were framed and organized into a survey document using Google Forms. Effort was taken to reduce mutual dependency between questions. The participating group was provided the set of the test specification qualities listed in Section 2.4 and requested to rate them based on their experience. The rating was on a Likert scale of 1–10, where 1 stands for "not relevant" and 10 for "extremely relevant". They were also encouraged to suggest other relevant qualities that are not listed in the questionnaire, but important to legacy systems. To better understand the needs and difficulties in developing quality test specifications, they were also asked to provide comments on the factors that help build quality test specifications and challenges. The survey was conducted online as a personalized survey where known responders (samples) were invited to take the survey. Responses were evaluated with outliers eliminated. The following are research questions that this study aims to answer.

■ What are the most concerning test qualities for legacy systems?
■ What are the factors affecting the creation of good test cases?
■ What other challenges hinder the development of good quality test specifications from an industry perspective?

The survey participants consist of 35 software engineers who have been working on legacy systems with a diverse set of outdated technologies including Mainframe, Smalltalk, and older Java applications. Table 2.1 shows the length of work experience of the survey participants. People with various years of experience, ranging from less than 5 years to more than 10 years, participated in the survey, which brings in various perspectives from different levels of proficiency. Totally, 84.4% of

Table 2.1   Work Experience of Survey Participants

Experience	Number of People	Percentage
Less than 5 years	6	15.6
5–10 years	14	40.6
More than 10 years	15	43.8
Total	35	100

the participants have more than 5 years of experience with software development and testing.

Table 2.2 shows the roles of the participants in the projects that they have participated. The roles include developer, tester, and manager/team lead. Totally, 71.5% of the participants are developers and testers.

Table 2.3 shows the age of the applications on which the participants have worked. It ranges from 10 years to more than 20 years, and 76.5% of the applications are more than 15 years old. These applications are employed in various domains, primarily finance, banking, insurance, telecommunication, energy utilities, retail, and health care, as shown in Table 2.4. They are used for various purposes such as automated control, semiconductor manufacturing, payroll management, and retirement management. The diversity of applications, domains, technologies, project roles, and professional backgrounds in the survey brings in unbiased insights and results with broader applicability.

Figure 2.1 shows the results of the survey ordered by importance based on the average of survey ratings. From the results, it is observed that more important qualities are harder to achieve due to the characteristics of legacy systems discussed in Section 2.3. Thus, the quality importance in the results can be viewed as the level of desire for particular qualities. The top three most desired qualities are *accuracy, reusability,* and *completeness.* Accuracy is most desired with 9.31 importance out of 10. According to the participants, test specifications with precisely defined expected outcome are highly desired for accuracy. However, it is difficult to achieve

Table 2.2   Roles of Survey Participants in Projects

Role	Number of People	Percentage
Developer	17	48.6
Tester	8	22.9
Manger/team lead	10	28.6
Total	35	100

Table 2.3   Age of Software on Which Participants Have Worked

Software Age	Number of People	Percentage
10–15 years	8	23.5
15–20 years	10	29.4
More than 20 years	17	47.1
Total	35	100

Table 2.4   Domains of Software on Which Participants Have Worked

Application Domains	Purpose	Number of People	Percentage
Finance, banking, and insurance	Payroll management	15	41.9
Telecommunication	Network inventory management	6	16.1
Energy utilities	Automated control	6	16.1
Health care and life science	Retirement management	2	6.5
Retail	Billing and customer handling	2	6.5
Others	Semiconductor manufacturing, monitoring systems	4	12.9
Total		35	100

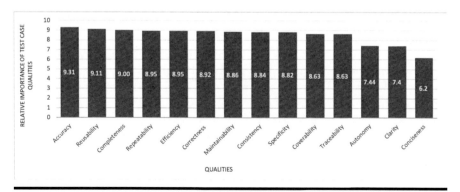

Figure 2.1   Test specification qualities by importance.

high accuracy due to the incomplete knowledge about the systems, which is the reason for the high desire. Reusability is ranked second with 9.11 importance. Per the participants' comments, developing test specifications for legacy systems is hard and time-consuming due to the characteristics discussed in Section 2.3, and reusable test specifications significantly help reduce testing efforts. One way to improve reusability is to make the use case of test specifications independent of input data, so that test specifications can be reused with different input data. A similar practice can be found in object-oriented programming where operations are defined with polymorphic parameters, which allow to pass an object of any type under the polymorphic type of the parameters in the class hierarchy. The third important quality is completeness with 9.00 importance. Completeness complements accuracy. Thus, as accuracy is rated highly and so is completeness as shown in the results. Also, the participants commented that higher completeness of test specifications leads to higher coverage of requirements.

Repeatability, efficiency, correctness, and *maintainability* are evaluated as the next group of important qualities. With the rating of 8.95 importance, repeatability is desired for checking the recurrence of the errors that have been discovered. Efficiency is rated as high as repeatability. The participants commented that efficiency is desired for not only efficient finding of errors but also efficient execution of test specifications, which helps identify missing implementation of requirements. Correctness is also rated as important because it has a direct impact on the quality of the system under test. The participants commented that developing correct test specifications is difficult due to the lack of the internal knowledge about the system. Maintainability is rated as desired for the evolution of the system. As the system evolves, test specifications that are related to the changes must also be updated. Maintainability also helps reduce the dependencies among test specifications, which promotes reusability as well.

Consistency, specificity, coverability, and *traceability* follow as the next desired qualities. However, their ratings are marginally less than the previous group. For consistency, which is rated as 8.84 of importance, the participants commented that the test specifications that are similar in purpose (e.g., user interfaces, functions) should be designed in a consistent manner. This helps identify repeated steps in test specifications and merge those specifications have a significant overlap, which reduces testing efforts and thus, improves testing efficiency. Specificity, which was also viewed as atomicity by some participants, is rated as 8.82 of importance. The participants commented that test specifications being specific to requirements help improve traceability to requirements. This is useful for identifying missing implementations of requirements. Specificity also promotes clear differentiation between positive and negative test specifications, which improves completeness as well. Coverability is rated as 8.63 of importance, which reflects the importance of code overage. The fact that this quality is rated relatively lower than other requirement-related qualities (e.g., completeness, specificity) reflects that the practitioners focus more on requirement-level testing rather than code-level testing. Traceability is also rated as 8.63 of importance. Traceability is necessary for improving specificity which is rated higher and thus, it is not surprising to see the high rating for traceability.

Autonomy, clarity, and *conciseness* are rated as less important than others. With respect to autonomy, the practitioners did not take the independency of test specifications as a desperate quality; although it is useful if exists. For clarity, the participants suggested making the concept more concrete as having step-wise details described in a simple manner rather than being understandable. Conciseness was taken as the least concerned quality and viewed as part of clarity by the participants. The participants commented that, if separated, clarity should be given more importance than conciseness as shown in the results. They also mentioned that the conciseness process might involve high complication (e.g., putting multiple steps into a single step), and it might depreciate understandability.

After all, the survey shows that the practitioners demand more qualities that are hard to achieve due to the characteristics of legacy systems discussed in Section 2.3. They are less concerned with qualities pertinent to maintenance and understanding. The ratings are marginally different and there is no outstanding quality in preference, which indicates close interrelations and dependencies among them. The qualities are also related to each other for mutual influence. The survey participants also suggested a few other qualities that they felt as highly desired for testing legacy systems.

- **Diversity:** Test specifications should be diverse enough to cover not only functionalities but also quality aspects such as performance and security with various types of input data and scenarios. For example, test should be carried out to understand the performance of the system in terms of normal, abnormal, and boundary conditions with both valid and invalid data. These conditions provide a gauge on the error handling capacity of the system.
- **Dissimilarity:** Test specifications should be considerably different from each other. This helps reduce the overhead of running the same or similar tests repeatedly and thus, improves efficiency. This quality is different from diversity in that it focuses on different purposes on the same or similar functions, while diversity is concerned with covering different aspects of functions and qualities. This quality proliferates diversity.
- **Usability:** Test specifications should be readily instantiated and executed. This includes defining executable steps for a specific functionality per the target requirement.
- **Preconditions:** Test specifications should include preconditions under which the test should be performed. Defining a precondition requires a clear understanding of system states as it determines the state in which the system should be before the test is carried out. If the precondition is not satisfied, the output of the test might not be guaranteed. However, preconditions are often neglected in practice due to the difficulty of acquiring information on the state-based behaviors of legacy systems.

■ *Change impact:* When a new or modified functionality is tested, the existing functionalities that are impacted by the change should also be tested to verify any side effects caused by the change. However, the survey participants acknowledge the difficulty in testing existing functionalities due to the difficulty in analyzing the dependencies of existing functionalities.

The survey also asked the participants to comment on the major factors that affect the development of quality test specifications. These factors can serve as the prerequisites for developing test specifications. Figure 2.2 shows the responses. Totally, 63.2% of the participants responded that requirements clarity (31.6%) and application knowledge (31.6%) are the most influencing factors. They acknowledge that a clear understanding of requirements is required to set up boundary conditions, scope, and expected outcome. The knowledge of application design, architecture, and behaviors is also highly desired in developing quality test specifications. The practitioners also advise to utilize existing test specifications to obtain the internal knowledge of the system.

With 21.1% responses, domain knowledge is identified as an important influencing factor in developing quality test specifications. However, they commented that domain knowledge alone is not sufficient and should be accompanied with the two aforementioned factors. Remaining 15.8% of responses advocate the understanding of user behaviors as testing is closely related to the users of the system.

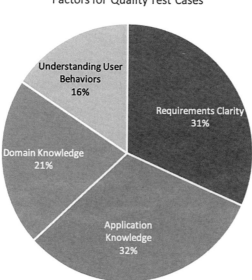

Figure 2.2   Factors for developing quality test specifications.

In fact, many errors are caused by users entering invalid data and misusing functions (e.g., not following the sequence of steps to perform a function) [15].

Apart from the characteristics discussed in Section 2.3, the survey participants also brought in two other concerns in maintaining and enhancing legacy systems based on their experiences, which should be taken into account in developing test specifications.

- ■ ***Data loss and recovery:*** As a legacy system has been through a long period of maintenance, the data stored or related to the system becomes vulnerable to loss. The organization operating the system should be prepared for data corruption and loss such as regular data backup or separating data from the system.
- ■ ***Data inconsistency:*** The data used in a legacy system needs to be maintained consistently. The users of the system use different tools to produce and manage data as the system evolves. These tools become obsolete over time and the data produced by the tools vary in format and type, which leads to data inconsistency. Also, the outdated tools are not capable of managing the increasing volume of data as the system evolves, which makes it difficult to leverage big data. Practitioners also noted that integrating legacy systems with modern data-handling tools is a challenge due to incompatibility.

In compilation of all feedback points, the results of the survey answer the search questions posed in Section 2.5 as follows.

- ■ What are the most concerning test qualities for legacy systems?
  Based on the study, accuracy was recognized as the most important quality and conciseness the least relevant. As per ratings from industry experts, the qualities were ordered based on their desirability.
- ■ What are the factors affecting the creation of good test cases?
  Application knowledge was identified to be the most important factor that aids in creation of test cases of good quality in legacy systems. Clear requirements, domain knowledge, and proper understanding of user behaviors are other relevant factors.
- ■ What other challenges hinder the development of good quality test specifications from an industry perspective?
  Data loss and recovery as well as data inconsistency due to application aging and obsoletion of tools and technology are identified as major challenges.

2.6 Threats to Validity

The main threats to validity in this work include population validity, content validity, and data extraction validity.

2.6.1 Population Validity

We discuss population validity in terms of application age and technology diversity.

- **Application age:** To derive meaningful results, it was important to have people who have been working with software applications older than at least 15 years participate in the survey. Newer applications may have some form of formal documentation or people who developed the system may still be available in the organization to provide information about the system. In such a scenario, the developer may be involved in a more informed development process and may not be facing the difficulties assumed in the survey. As described in Section 2.5, 76.5% of the participants represent the valid category of application age greater than 15 years.
- **Technology diversity:** We tried to include in the survey people from diverse technology and domain backgrounds as well as varying experiences. This measure helped us come up with more generalized results applicable to a broad range of legacy systems. However, there may be other legacy technologies that are not considered in the survey.

2.6.2 Content Validity

Prior to preparing the survey, we conducted a literature review to identify the qualities that are perceived as important in test specifications for legacy applications. The identified qualities are shortlisted to avoid overlaps. It is possible that the survey may have missed certain qualities that are important from the perspective of this study. To mitigate this issue, we provided a space in the survey form for participants to suggest other qualities that they feel are important other than the ones listed in the survey.

2.6.3 Data Extraction Validity

The rating of qualities was calculated by averaging out the responses from survey takers. During this process, the responses that lay well outside the range of majority opinions were eliminated. Due to this, some valuable responses may have been lost and their inclusion might have produced slightly different results.

2.7 Conclusion

Legacy systems are widely used across domains including banking, finance, telecom, energy utilities, and health care. They need to be updated with new features and functionalities to keep up with growing needs in their respective domains. The lack of documentation and technical expertise makes it difficult to maintain and enhance legacy systems. The survey reveals that the most concerned test quality for legacy systems is accuracy. Other surveyed qualities are also given high ratings with marginal differences, which indicate close interrelations and dependencies among them. Apart from the surveyed qualities, the survey participants suggested other qualities based on their experiences, which are also found to be related to the surveyed qualities. The survey also identified major factors that affect the development of quality test specifications. These factors can serve as the prerequisites for developing test specifications. The professionals who participated in the survey identified data loss and recovery and data inconsistency as major difficulties in maintaining and enhancing legacy systems based on their experiences, which should be taken into account in developing test specifications. The definition of legacy systems has been extended to include web-based systems to be migrated to the cloud [16]. A further study can be conducted to better understand quality concerns for test specifications of these systems. We also plan to further examine the importance of the qualities that are suggested by the participants of this survey.

References

1. A. Dedeke, "Improving Legacy-System Sustainability: A Systematic Approach," *IT Professional*, vol. 14, no. 1, pp. 38–43, 2012.
2. N. H. Weiderman, J. K. Bergey, D. B. Smith and S. R. Tilley, "Approaches to Legacy System Evolution," Carnegie Mellon University, 1997.
3. S. Comella-Dorda, K. Wallnau, R. C. Seacord and J. Robert, "A Survey of Legacy System Modernization Approaches," Carnegie Mellon University, 2000.
4. J. Lee, S. Kang and D. Lee, "A Survey on Software Testing Practices," *IET Software*, vol. 6, no. 3, pp. 275–282, 2012.
5. I. Jovanovikj, V. Narasimhan, G. Engels and S. Sauer, "Context-Specific Quality Evaluation of Test Cases," in the Proceedings of the 6th International Conference on Model-Driven Engineering and Software Development, 2018. https://www.scitepress.org/Papers/2018/67244/67244.pdf
6. S. M. Hussain, S. N. Bhatti and M. F. U. Rasool, "Legacy System and Ways of Its Evolution," in the Proceedings of International Conference on Communication Technologies, Rawalpindi, Pakistan, 2017.
7. D. Nirmala and T. LathaMaheswari, "Automated Testcase Generation for Software Quality Assurance," in the Proceedings of the 10th International Conference on Intelligent Systems and Control, Coimbatore, India, 2016.
8. A. Adlemo, H. Tan and V. Tarasov, "Test Case Quality as Perceived in Sweden," in the Proceedings of the 5th International Workshop on Requirements Engineering and Testing, 2018. DOI: 10.1145/3195538.3195541.
9. C. Holland and B. Light, "A Critical Success Factors Model for ERP Implementation," *IEEE Software*, vol. 16, no. 3, pp. 30–36, 1999.
10. K. Bennett, "Legacy Systems: Coping with Success," *IEEE Software*, vol. 12, no. 1, pp. 19–23, 1996.
11. J. Bisbal, D. Lawless, W. Bing and J. Grimson, "Legacy Information Systems: Issues and Directions," *IEEE Software*, vol. 16, no. 5, pp. 103–111, 1999.

12. I. S. 610.12-1990, *IEEE Standard Glossary of Software Engineering Terminology*, IEEE, New York, NY, 1990.

13. B. Zeiss, D. Vega, I. Schieferdecker, H. Neukirchen and J. Grabowski, "Applying the ISO 9126 Quality Model to Test Specifications – Exemplified for TTCN-3 Test Specifications," in Software Engineering, Hamburg, 2007.

14. M. Atmadja, Z. Shuai and Richard, "Chapter 3: Testing," in *A Fresh Graduate's Guide to Software Development Tools and Technologies, School of Computing*, National University of Singapore, Singapore, 2011, pp. 1–21.

15. S. Andrica, "Testing Software Systems Against Realistic User Errors," ÉCOLE POLYTECHNIQUE FÉDÉRALE DE LAUSANNE, 2014.

16. M. F. Gholami, F. Daneshgar, G. Beydoun and F. Rabhi, "Challenges in Migrating Legacy Software Systems to The Cloud – An Empirical Study," *Information Systems*, vol. 67, pp. 100–113, 2017.

17. J. Crotty and I. Horrocks, "Managing Legacy System Costs: A Case Study of a Meta-assessment Model to Identify Solutions in a Large Financial Services Company," *Applied Computing and Informatics*, vol. 13, no. 2, pp. 175–183, 2017.

18. T. Punter, M. Ciolkowski, B. Freimut and I. John, "Conducting On-line Surveys in Software Engineering," in the Proceedings of the International Symposium on Empirical Software Engineering, Rome, Italy, 2003.

Chapter 3

Whom Should I Talk To?: And How That Can Affect My Work

Subhajit Datta[a]

[a]*Singapore Management University, Singapore*

Contents

3.1   Introduction...76
3.2   Methodology..77
    3.2.1   General Approach...77
    3.2.2   Generating Defect Discussion Network.............................78
    3.2.3   Model Variables...78
        3.2.3.1   Dependent Variable...79
        3.2.3.2   Independent Variables......................................79
        3.2.3.3   Control Variables..79
    3.2.4   Choice of a Modelling Paradigm......................................80
3.3   Results...80
3.4   Discussion...82
    3.4.1   Validating the Hypotheses...82
    3.4.2   Implications and Utility of Our Results.............................82
    3.4.3   Towards a Novel Perspective...83
3.5   Threats to Validity and Future Work..83
3.6   Related Work...84
3.7   Conclusion..85
Notes 85
References 86

3.1 Introduction

Large scale software development has gone increasingly global in the past two decades. Along with it, the essence of collaboration in distributed development has gained wider recognition. Ever since Raymond canonized Linus' Law, we have come to believe that large and distributed teams can collectively build complex software, successfully [1]. A class of software programs meant to facilitate collaboration – the so-called collaborative development environment or CDEs – are making it easier for developers to connect and build software conjointly.

The argument in favor of collaboration is well established and repeated almost to the point of being clichéd. Collaboration enables the exchange of ideas, perception, and knowledge; fostering an environment of originality and skill building. Collaboration is also very much part of our *zeitgeist:* "… the lone genius is a myth that has outlived its usefulness. Fortunately, a more truthful model is emerging: the creative network, as with the crowd-sourced Wikipedia or the writer's room at 'The Daily Show' … " [2]. However, there is a contrarian view too, voiced by none other than Dijkstra: "… thanks to the greatly improved possibility of communication, we overrate its importance. Even stronger, we underrate the importance of isolation. Thanks to my isolation, I would do things differently than people subjected to the standard pressures of conformity. I was a free man"[1]. Dijkstra, in all his prescience, was pointing out something subtle and unsettling. These days, as we measure ourselves mainly by the "likes" we collect from our social network, are we submitting to the very uniformity that collaboration beckons us to break out from?

Given the deeply social nature of large scale and distributed software development, how developers connect to their peers significantly impacts individual and team outcomes [3]. As software systems become increasingly complex, such effects have deep consequences for all stakeholders in the development ecosystem. Thus, an understanding of these situations in light of empirical evidence is essential for optimal tuning of development processes. Developers today have a wide array of tools and technologies that help them work together. But given this range of choice, whom do we choose to connect to in our peer group? And how does this choice influence the outcome of what we do? We formalize these concerns into the following research question we will examine in this chapter:

Does uniformity of developer interaction relate to more effective team outcome in large scale software development?

Software making has long been rent by duality at more than one level:

> whether it is art versus science, or craft versus engineering [4]. To repeatably deliver industrial artefacts of consistent quality, a high degree of regularity in the production process is required. This is best achieved by teams that are predictable in their outcomes. On the other hand, teams are often goaded to "think out of the box" to bring about the next disruptive technology.

This tension between uniformity and diversity confronts large scale software making today, as larger and more distributed teams are assembled. With this perspective we distil the research question into the following hypotheses. In each case we first state the *alternative hypothesis*, followed by the corresponding *null hypothesis*.

- X_a: *Higher collaboration among members of a team relates to lower number of defects in the team's work products.* The corresponding null hypothesis X_0 is that there is no relationship between collaboration among members of a team and the number of defects in the team's work products.
- Y_a: *Higher uniformity in interaction among members of a team relates to lower number of defects in the team's work products.* The corresponding null hypothesis Y_0 is that there is no relationship between interaction among members of a team and the team's work products.

We are assuming that lower number of defects raised on a team's development activities signifies a better outcome, which aligns with established views on software quality [5]. As we discuss in the Methodology section, we measure collaboration and uniformity using standard metrics.

To test the above hypotheses, we study a set of 389 projects, across 3,26,249 defects, involving 2,02,972 developers, who raise 12,50,841 comments in the GNOME suite of projects[2]. GNOME is an open-source desktop environment, meant for computers running Unix-like operating systems. The GNOME system is developed by The GNOME Project, which is part of the GNU Project. Participants in The GNOME Project are both volunteers as well as paid contributors[3]. The data set we study in this chapter was extracted from the development environment, curated, and made available by the organizers of the Working Conference on Mining Software Repositories (MSR) as a part of a mining challenge[4].

3.2 Methodology

In the following subsections, we discuss the components of the methodology used in this study.

3.2.1 General Approach

The major steps in our methodology involve collecting the data set (from the source mentioned above), cleaning, and filtering it for errors and omissions, generating a defect discussion network (as explained below), extracting parameters for our statistical models, testing hypotheses around the relations between the independent variables and the dependent variable in presence of the control variables (as identified below) and deriving insights based on the results.

3.2.2 Generating Defect Discussion Network

In large scale collaborative development, developer discussion around work items is a key mechanism for collective completion of tasks. In our study setting, defects were discussed widely by developers. As depicted in Figure 3.1, we extracted a *defect discussion network* (DDN) by folding the bipartite network of developers commenting on defects to a network of developers based on instances of developers co-commenting on defects. The vertices of a DDN are developers, and two developers are connected by an edge, if both the developers have commented on one or more defects in a particular project. We leverage the network paradigm for our study in tune with its wide usage in the study of collaborative enterprise [6].

3.2.3 Model Variables

We considered the following variables in our model with a view to understanding the effects of the independent variables (collaboration and uniformity of interaction) on the dependent variable (number of defects), after accounting for the peripheral effects of the control variables (factors which are known to influence defect counts in a project). As illustrated in the Results and Discussion section, we build two sets of models: the *base model* specifies the effects of the control variables on the dependent variable, while in the subsequent *refined model* we additionally include the independent variables. By comparing the parameters of the base and the refined models, we seek to isolate the incremental influences of the independent variables on the dependent variable, over and above those of the control variables on the latter. Each of the base and refined models are developed using data from

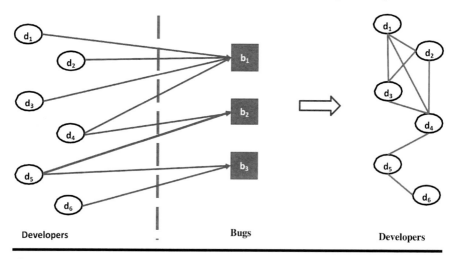

Figure 3.1 **Developers commenting on defects (left) and the corresponding defect discussion network (DDN) extracted from instances of developers co-commenting on defects (right).**

389 projects from the data set. Each of these projects had at least three comments by developers, to enable the generation of non-trivial DDN for each project.

3.2.3.1 Dependent Variables

The output of software development teams can be measured in different ways. As widely studied in the literature, the number of defects resolved by a team is an important parameter from the project governance perspective, as well as a reflection of team dynamics [7]. We consider the total number of defects (**DefectCount**) for a project as our dependent variable.

3.2.3.2 Independent Variables

Newman has shown how social networks differ from other networks in two important ways – assortative mixing and non-trivial clustering [8]. Assortative mixing in networks is a reflection on the tendency for vertices "to be connected to other vertices that are like (or unlike) them in some way" [9]. Our intuitive sense for *homophily* – the propensity of individuals to associate with *similar* others – is captured in the notion of assortative mixing [10]. In a network, the clustering coefficient (C_v) for a vertex v is expressed as: If v has a degree of k_v, that is there are k_v vertices directly linked to v, the maximum number of edges between these k_v vertices is k_v *choose* 2 or $k_v \times (k_v - 1)/2$. If the *actual* number of such edges existing is N_v, then $C_v = 2 \times N_v \div k_v \times (k_v - 1)$. Thus, the clustering coefficient of a vertex is the ratio of the actual number of edges existing between its neighbors and maximum number of such edges that can exist [11]. For an entire network, the clustering coefficient is the average clustering coefficient of all its vertices.

Since clustering reflects the level of collaboration in a team [5], we consider **Collaboration** – as measured by the clustering coefficient of each project's DDN – as an independent variable in our model. Assortative mixing in a network can be calculated by measuring the degree correlation of vertices [9, 12]. Using the method outlined in [13] we calculated the Pearson correlation coefficient between the degrees of pairs of vertices at the ends of the edges of each project's DDN. Evidently, higher assortativity indicates connection between developers who are more alike. We calculate **Uniformity** as each DDN's assortativity, and include it as another independent variable in our model.

3.2.3.3 Control Variables

To control for some of the important peripheral effects on the dependent variable [5], we use the following control variables in our model:

- **Connection**: The average number of edges per vertex, that is, the average degree is a reflection of a network's interconnectedness [14]. For a network with V vertices and E edges, average degree can be calculated as $2 \times E/V$. We take the average degree of each project's DDN as the *Connection* control variable.

- ***Distance***: In a network, whether and how two vertices i and j communicate with one another is influenced by the length of the *shortest* path l_{ij} between them [14]. The average of l_{ij} over all pairs of vertices is the average separation of each project's DDN; it is taken as the *Distance* control variable.
- ***Importance***: To account for the level of importance of vertices in each project's DDN, we take the eigenvector centralization [15] as the *Importance* control variable.
- ***Focus***: The number of defects in a particular project can be influenced by how focused the developers are on the defect resolution process. Accordingly, we include *Focus* as a control variable, which is calculated as the elapsed time between the first and last comment on a defect in each project.
- ***InterestSpan***: To capture the effect of the breadth of knowledge of developers, we calculate the *InterestSpan* control variable as the median number of comments on defects made by developers in each project.

3.2.4 Choice of a Modeling Paradigm

As identified above, our dependent variable is a *count* of defects for each project. Thus, Poisson regression was first considered as a modeling paradigm. Overdispersion, which involves the violation of the underlying assumption of the equality of variance and mean of a Poisson distribution, represents a major threat to the validity of Poisson regression [16]. As this threat was found to be notable in our context, multiple linear regressions were deemed to be a more suitable modeling paradigm. The assumptions underlying multiple linear regressions are linearity, normality, and homoscedasticity of the residuals, and absence of multicollinearity between the independent variables. The residual properties were verified using histogram, P-P plot and scatter plot of the standardized residuals. The *Connection* control variable was transformed by taking its square root, for closer fit to normality. The correlations among the independent variables were low (around −0.37). To further verify whether multicollinearity between variables artificially altered the significance of the overall regression and the stability of the regression coefficients, we calculated the variance inflation factor (VIF) for each variable. For all the variables, the VIF was found to be below the recommended upper limit [17]. On the basis of the above discussion, it can be concluded that the assumptions of linear multiple regression hold within permissible limits in our context. So it is appropriate to use multiple linear regressions as the modeling paradigm [17].

3.3 Results

Table 3.1 describes the base and refined models. In columns I and II of Table 3.1, the regression coefficients of each variable in the model are presented, with the corresponding standard errors in parentheses below the coefficients. As specified

Table 3.1   Results of Regression for the Effects on Defect Count

	I *Base Model*	*II* *Refined Model*
Intercept	1.86 (1.57)	2.02 (1.54)
Control Variables		
Focus	1.72×10^{-4}*** (4.21×10^{-5})	1.72×10^{-4}*** (4.15×10^{-5})
Connection	4.99*** (0.326)	5.24*** (0.363)
Distance	21*** (4.31)	15.7*** (4.51)
Importance	−37.6*** (7.62)	−0.195* (8.97)
InterestSpan	5.21×10^{-7}*** (1.27×10^{-7})	5.09×10^{-7}*** 1.25×10^{-7}
Independent Variables		
Collaboration		−13.8** (4.3)
Uniformity		8.53* (3.9)
	Model Parameters	**Model Parameters**
N	389	389
R^2	0.632	0.645
Df	383	381
F	132	98.8
P	< 0.001	< 0.001

***$p \leq 0.0001$

**$p \leq 0.001$

*$p \leq 0.01$

.$p \leq 0.05$

in the table caption, superscripts of the coefficients indicate ranges of their respective *p* values. The *p* values are calculated using the t-statistic and the Student's t-distribution. In the table's lower section, parameters of each model are given, where N is the number of data points used in the model development. In our context N is the number of projects considered in the analysis. R^2 denotes the coefficient of determination, which is the ratio of the regression sum of squares to the total sum of squares. R^2 indicates the goodness-of-fit of the regression model as a proportion of variability in the data set that is accounted for by the model. The degrees of freedom are denoted by *df.* The Fisher F-statistic is represented by F – the ratio of the variance in the data explained by the linear model divided by the variance unexplained by the model. The *p* value for each model is calculated using the F-statistic and the F-distribution, and it indicates the overall statistical significance of the model. For each coefficient as well as the overall regression models, we conclude the corresponding result is statistically significant on the basis of null hypothesis significance testing if $p \leq$ *level of significance*. The specific level(s) of significance we have considered are mentioned in the caption of Table 3.1.

3.4 Discussion

3.4.1 Validating the Hypotheses

With reference to Table 3.1, we observe that both models are statistically significant overall, and the goodness of fit improves between the base and refined models from 63% to about 65%. This shows that addition of the independent variables over and above the control variables in the refined vis-a-vis the base model enhances the goodness-of-fit by about 3%. We also note that the effects of both the independent variables on the dependent variable are statistically significant. *Thus, higher level of collaboration among team members relate to lower number of defects for a team, and higher level of uniformity among interacting developers is associated with higher number of defects for a team.* On the basis of these results, we find statistically significant evidence to reject the null hypotheses X_0 and Y_0 corresponding to X_a and Y_a respectively (as defined in Section 3.1): the level of collaboration as well as the extent of uniformity in the interaction between developers *does* relate to the number of defects in teams' work products. The association between the level of collaboration and the number of defects is as hypothesized in X_a; however, the relation between the extent of uniformity and the number of defects is found to be contrary to what was hypothesized in Y_a.

3.4.2 Implications and Utility of Our Results

We find evidence that more collaboration in a project team relates to fewer defects. This is not unexpected. As developers collaborate more, there is higher permeation of knowledge and skills in the project ecosystem, which is reflected in higher effectiveness of the team as they develop artefacts with fewer defects. But higher uniformity, that is higher tendency for developers to interact with peers who are similar to themselves, is found to be associated with higher number of defects. In software projects, collaboration largely comes out of the project mandate to deliver quality artefacts within the constraints of time and budget. However, the choice of whom to interact with by way of collaboration is largely left to the discretion of individual developers. As we discussed earlier, the tendency for individuals to interact with peers who are similar to themselves has roots in the deeply social trait of homophily. Our results indicate that this propensity may actually be detrimental to project outcome. So, merely facilitating collaboration through tools, processes, and words of encouragement or enforcing it through management diktat may not be enough to actuate developers to reach out to a diverse group of peers.

These observations have notable implications for effective assembly and governance of teams in large scale software development. At the individual level, developers can consciously focus on cultivating a circle of peers with skills and interests different from theirs. Our results emphasize the need for project managers to allocate tasks such that team members with diverse backgrounds have an opportunity of working together. As organizations grapple with the challenges of a globally distributed workforce, our results can offer a novel perspective into the benefits of harnessing the inherent lack of uniformity in such scenarios.

3.4.3 Towards a Novel Perspective

Given the interactional nature of software development, the quest for "gelled" teams has long been of interest to practitioners [18]. While it is easy to recognize such a team by its effective functioning, an approach toward repeatably assembling such teams in the flux of enterprise software development is elusive. Great leaders in society and government bring their original strategies to team-making; these often run contrary to conventional wisdom, yet they can yield successful outcomes. A notable example is Abraham Lincoln's "team of rivals" [19]. Within the limitations identified in the next section, our results introduce a novel perspective on the counter-intuitive interplay of factors in the collective enterprise of large scale software development. The size and complexity of today's software systems make it imperative for individual developers to collaborate. *But for collaboration to be most effective, developers need to interact with those unlike themselves.* In a world where individuals are increasingly trapped in their own "echo chambers", this message opens a new perspective on how software developers need to work together to be most effective [20].

3.5 Threats to Validity and Future Work

Before we identify specific threats, we need to recognize that this is an observational study rather than a controlled experiment. Thus, correlation *does not* imply causation. The ideal setting to determine if higher collaboration and higher uniformity indeed cause lower defects would be an experiment where developers are separated into control and treatment groups, and the effects of the independent variables on the dependent variable are measured. However, such a study is difficult to design and execute in a real world software development scenario. In our future work, we plan to analyze how the results from this study can be extended to identify causal effects.

Construct validity – relates to whether the variables are measured correctly. As mentioned earlier, the network paradigm is widely used for studying team dynamics. Our independent variables are measured using established network metrics. The measurement of control variables are also grounded in standard software engineering practices. However, we recognize that there may be alternative ways to measure the model variables [21]. The way we have constructed the team network is one among several approaches to building similar networks [22, 23].

Internal validity – ensures that a study is free from systematic errors and biases. As the GNOME repository is our only source of the data, common issues affecting internal validity such as mortality and maturation do not arise in this case. However, a key influence on internal validity is the extent to which the development team used the defect resolution platform, which was the original source of the data.

External validity – is concerned with generalizability of the results. Our study ranges across the development cycle of multiple projects in a single suite of products. In our future work, we plan to extend this study to other suits of products and platforms, to establish whether our results are generalizable. However, as demonstrated in existing literature, useful insights can be drawn from observational studies on limited number of subjects [22, 24, 25]. So, even as we do not claim our results to be generalizable as yet, we believe they can offer insights on the study objective.

Reliability – of a study is established when the results are reproducible. From the discussion in earlier sections, it is evident that there are no particular interventions in the extraction and processing of data. Once the data is extracted, we use standard statistical technique for analysis. Assuming our data source is accessible, our results can be easily reproduced.

3.6 Related Work

We have cited relevant literature as we were developing our hypotheses and models. We briefly outline some additional related work here. In recent times several empirical studies have been based on developer interaction. In contrast to previous results, Wolf et al. have established that distance between developers does not really matter in task completion for development on the Jazz platform [24]. Kwan et al. report a surprising result from their study of the Rational Team Concert development project: higher socio-technical congruence leads to lower build success rate [26]. It has been found that individuals who are more central within a team's communication network performed better, but those more central within the network for the whole project performed worse [23]. Cataldo and Herbsleb have studied communication networks in geographically distributed software development to conclude that core developers in the project are highly productive [21]. They suggest that collaborative development environments should make it easy for such people to easily move between technical work and communication. Hinds and McGrath studied 33 R&D teams and found some support for the perception that loosely-coupled distributed teams are better able to coordinate their work [27]. Costa et al. have studied the scale and temporal evolution of coordination requirements in five large scale distributed projects and found the coordination requirements to be quite volatile and are significantly different for different team members [28]. In recent studies, we have examined how evolutionary trends manifest in collaborative peer review processes [29] and the relation between influence, information, and team outcomes in large scale software development [30]; we find evidence of the effects of connection and clustering on software quality in another study of a large scale industrial software system [5].

3.7 Conclusion

In this chapter, we addressed a research question around whether and how the uniformity of developer interaction relates to team outcomes in large scale software development. We validated two hypotheses derived from the research question using data from a set of nearly 400 real-world projects. For each project we extracted a DDN from instances of developers co-commenting on defects, as they collectively worked toward their resolution. Using a set of established metrics, we captured the notions of developer collaboration and uniformity of their interaction, along with other peripheral factors such as focus, connection, distance, importance, and interest span. As outcome measure, we considered the number of defects raised against the work products of each team. We found statistically significant evidence that higher levels of collaboration relates to lower number of defects whereas higher uniformity in developer interaction is associated with more defects. These results highlight the need for a deeper understanding of the nuances of developer communication in software development ecosystems, at individual and collective levels. As globally distributed teams become the norm in large scale software development, our results can enable more effective team interaction, as well as facilitate higher individual enrichment from the interactive processes of software building. To return to the metaphorical question that is in the title of this chapter, whom I choose to talk to can indeed influence the outcome of my work in subtle and significant ways!

Notes

1. An Interview with Edsger W. Dijkstra – http://bit.ly/1MIdriW
2. https://www.gnome.org/
3. https://en.wikipedia.org/wiki/GNOME
4. http://2009.msrconf.org/challenge/msrchallengedata.html

References

1. Raymond, E. S. *The Cathedral and the Bazaar: Musings on Linux and Open Source by an Accidental Revolutionary.* Boston, MA: O'Reilly, 2001.
2. Shenk, J. W. Opinion | The End of 'Genius'. *The New York Times* (Jul. 2014).
3. Datta, S., Sindhgatta, R., and Sengupta, B. Talk versus work: characteristics of developer collaboration on the jazz platform. In *Proceedings of the ACM International Conference on Object Oriented Programming Systems Languages and Applications* (New York, NY, USA, 2012), OOPSLA '12, ACM, pp. 655–668.
4. Glass, R. L. *Software Creativity 2.0.* developer.* Books, Nov. 2006.
5. Datta, S. How does developer interaction relate to software quality? An examination of product development data. *Empirical Software Engineering 23*, 3 (June 2018), 1153–1187.
6. Guimera, R., Uzzi, B., Spiro, J., and Amaral, L. A. N. Team assembly mechanisms determine collaboration network structure and team performance. *Science (New York, N.Y.) 308*, 5722 (Apr. 2005), 697–702. PMID: 15860629.
7. Koru, A. G., AND Liu, H. Building defect prediction models in practice. *IEEE Softw. 22*, 6 (Nov. 2005), 23–29.
8. Newman, M. E. J., and Park, J. Why social networks are different from other types of networks. *cond-mat/0305612* (May 2003). *Phys. Rev. E. 68* (2003), 036122.
9. Newman, M. E. J. Assortative mixing in networks. *arXiv:cond- mat/0205405* (May 2002). *Phys. Rev. Lett. 89* (2002), 208701.
10. McPherson, M., Smith-Lovin, L., AND Cook, J. M. Birds of a feather: Homophily in social networks. *Annual Review of Sociology 27*, 1 (2001), 415–444.
11. Albert, R., and Barabasi, A. Statistical mechanics of complex networks. *cond-mat/ 0106096* (June 2001). *Reviews of Modern Physics 74* (2002), 47.
12. Newman, M. E. J. Mixing patterns in networks. *arXiv:cond- mat/0209450* (Sept. 2002). *Phys. Rev. E. 67* (2003), 026126.
13. Newman, M. E. J. The structure and function of complex networks. *cond-mat/0303516* (Mar. 2003). *SIAM Review 45* (2003), 167–256.
14. Barabasi, A. L., Jeong, H., Neda, Z., Ravasz, E., Schubert, A., and Vicsek, T. Evolution of the social network of scientific collaborations. *cond-mat/0104162* (Apr. 2001). *Physica A 311*, 3–4 (2002), 590–614.
15. Batagelj, V., and Mrvar, A. Introduction to social network methods. http://vlado.fmf. uni-lj.si/pub/networks/pajek/, 2009.
16. Barron, D. The analysis of count data: Overdispersion and autocorrelation. *Sociological Methodology 22* (1992), 179–220.
17. Tabachnick, B., and Fidell, L. *Using Multivariate Statistics.* Boston: Pearson Education, 2007.
18. Humphrey, W. S. Introduction to the team software process. *SEI Series in Software Engineering* (1999).
19. GOODWIN, D. K. *Team of Rivals: The Political Genius of Abraham Lincoln.* London: Penguin UK, Feb. 2009.
20. Quattrociocchi, W., Scala, A., and Sunstein, C. R. Echo Chambers on Facebook. SSRN Scholarly Paper ID 2795110, Social Science Research Network, Rochester, NY, June 2016.
21. Cataldo, M., and Herbsleb, J. D. Communication networks in geographically distributed software development. In *Proceedings of the 2008 ACM Conference on Computer Supported Cooperative Work* (New York, NY, USA, 2008), CSCW '08, ACM, p. 579–588.

22. Bird, C., Nagappan, N., Devanbu, P., Gall, H., and Murphy, B. Putting it all together: Using socio-technical networks to predict failures. In *Proceedings of the 17th International Symposium on Software Reliability Engineering* (2009), IEEE Computer Society.

23. Ehrlich, K., and Cataldo, M. All-for-one and one-for-all?: A multi-level analysis of communication patterns and individual performance in geographically distributed software development. In *Proceedings of the ACM 2012 Conference on Computer Supported Cooperative Work* (New York, NY, USA, 2012), CSCW '12, ACM, pp. 945–954.

24. Wolf, T., Nguyen, T., and Damian, D. Does distance still matter? *Softw. Process 13*, 6 (2008), 493–510.

25. Wolf, T., Schroter, A., Damian, D., and Nguyen, T. Predicting build failures using social network analysis on developer communication. In *Proceedings of the 31st International Conference on Software Engineering* (Washington, DC, USA, 2009), ICSE '09, IEEE Computer Society, pp. 1–11.

26. Kwan, I., Schroter, A., and Damian, D. Does socio-technical congruence have an effect on software build success? A study of coordination in a software project. *IEEE Transactions on Software Engineering 37*, 3 (2011), 307–324.

27. Hinds, P., and McGrath, C. Structures that work: social structure, work structure and coordination ease in geographically distributed teams. In *Proceedings of the 2006 20th Anniversary Conference on Computer Supported Cooperative Work* (Banff, Alberta, Canada, 2006), ACM, pp. 343–352.

28. Costa, J. M., Cataldo, M., and de Souza, C. R. The scale and evolution of coordination needs in large-scale distributed projects: implications for the future generation of collaborative tools. In *Proceedings of the 2011 Annual Conference on Human Factors in Computing Systems* (New York, NY, USA, 2011), CHI '11, ACM, pp. 3151–3160.

29. Datta, S., and Sarkar, P. Evolutionary trends in the collaborative review process of a large software system. In *Proceedings of the 12th Innovations on Software Engineering Conference (formerly known as India Software Engineering Conference), ISEC 2019* (Pune, India, February 14–16, 2019), R. Naik, S. Sarkar, T. T. Hildebrandt, A. Kumar, and R. Sharma, Eds., ACM, pp. 22:1–22:5.

30. Datta, S. Influence, information and team outcomes in large scale software development. In *26th Asia-Pacific Software Engineering Conference, APSEC 2019* (Putrajaya, Malaysia, December 2–5, 2019), IEEE, pp. 402–409.

Chapter 4

Software Project Management: Facts versus Beliefs and Practice

Lawrence Peters[a]

[a]*Universidad Politecnica de Madrid, Spain, and Iowa State University, Iowa*

Contents

4.1 Background...89
4.2 Capable Software Project Management 91
4.3 Hard versus Soft Management Skills...93
4.4 Hard versus Soft Skills 94
4.5 Changing the Current Situation..95
4.6 A Multicultural Management Practice 95
4.7 Managing Multiple Generations 96
4.8 What is a Successful Software Engineering Team?.......................97
4.9 Conclusion 98
References 98

4.1 Background

Computer programming started as a solitary endeavor in which a few people wrote code that had a relatively narrow focus. As hardware became more capable and reliable, the software being written became more pervasive in industry requiring larger groups of software developers. At some point, it was determined that these larger groups needed people to manage them. Unlike writing code, the practice of

89

software project management does not have a syntax and semantics or a beginning and a definable end by which software project managers are guided through a process to success. In most cases, the software project manager does not know if they got it right until it is too late to correct what they may have gotten wrong. Perhaps due to the fact that training in software project manager is not required in most software engineering curricula, software engineers are left to their own devices to figure out how to manage a software project. This has led to the institutionalization of a number of ineffective, counter-productive practices collectively referred to as anti-patterns [1]. These are actions that are taken which were believed to be beneficial but which had the opposite effect. Many of these seem reasonable but do not stand the test of measurement or experimentation regarding their effectiveness. A good example is the practice of splitting the time of a software engineer between two or more projects. This might involve having the software engineer spend 50% of their time on one project and 50% on another. The underlying problem with this practice is that it treats the software engineer like some sort of mechanical resource which can be practically switched from working on one thing then another. But even physical devices (e.g., a hard drive) require some time to spin down or up when switching tasks. Anyone who has looked closely at this practice has found that there is a loss in productivity in that what appears to be 50% + 50% adds up to 60% or less. This is because people behave like a slow hard drive. They take time to get up to speed on one effort, then spin down on that effort, then spin up on the second effort and so on and so forth. This task switching results in a loss of productivity and has the effect of increasing cost, by increasing the time to complete the tasks involved and/or forcing the hiring of additional personnel to make up for the lost productivity. Practices of this type constitute hidden impediments to success leaving researchers to conclude other factors caused the demise of a software project including using the Waterfall Lifecycle, a difficult relationship with the client, and other scapegoated factors while missing the real cause(s) of failure. A study of failed projects conducted by IBM [2] revealed that 53% of the failures were attributable to, "Poor Management" while 3% could be attributed to "technical challenges." The poor management factor may seem non-specific but that is only because there are so many ways in which the software project manager can negatively affect a project. One author [1] noted nearly 100 different negative actions the software project manager could take which he labeled, "anti-patterns." He also speculated that the number of negative actions may actually be higher because software project managers seem to be able to create new ways to sabotage a project without realizing the negative effect they are having. How can this be? In code development, the last 50 years have produced programming languages, design and analysis methods and software tools which have greatly improved the quality of delivered systems. Unlike the practice of software project management, the development of executable code is dealing with a relatively stable target which analysis tools, testing and the latest programming practices can improve the level of quality present in delivered systems. The management created environment within which these improved development practices are employed does not have the benefit of analysis tools with which to evaluate the quality of that environment. This means that quality software may be developed, in a sense, in spite of "poor management" but less often than an environment of capable management. What is a capable management environment like?

4.2 Capable Software Project Management

Throughout the practice of software engineering, there has been an underlying presumption that technology in many forms (e.g., programming languages, development methods) will solve software engineering challenges. This philosophy is exemplified in the common practice [3] of installing the best software engineer in a group as a software project manager. The imprudence of this practice has been studied and written about for several decades [4]. What is missed in engaging in this practice is the fact that taking a high performing individual out of the role of developing software has several negative side effects:

- The productivity of the group is reduced because one of its best performers is now dealing with personnel issues, participating in management meetings, tracking budget and project progress and other non-code development activities.
- Successful software teams are viewed as having achieved some skill that senior management would like to see spread throughout the software development organization. So successful teams are often broken up with members being assigned to various other projects in the hopes that their skill and success will spread. It turns out the experience and data are quite the opposite [5, 6]. These successful teams are successful, in part, because:
  - They are mutually supportive, helping each other in order to achieve success as a group rather than each member trying to outdo the others in order to be recognized for a promotion or salary increase.
  - The team members are well aware of the skills of other team members. When they run into a coding problem have difficulty solving, they know which of their colleagues to seek help from based on this knowledge. That streamlines problem solving making the team highly productive.
- Unless they are a very unique individual, because they are so skilled, they may not be very patient with those less skilled than they are causing the poorer performing to become demotivated. If they leave the project as a result, it could increase project cost by as much as 60% [7]. A better choice would be someone with great interpersonal skills and planning capabilities who is not the best performer in the group. This increases the overall productivity of the group while putting someone who is challenged by having to produce code but can plan, schedule, control, staff and motivate into a position where they will excel.
- They may be a poor manager because their proficiency in developing code demonstrated that coding is something they really liked doing. It seemed

natural to them. Pulling them away from that leaves them in a cognitive dissonance situation. That is, the increase in pay when promoted into management is welcomed but having to deal with a spectrum of issues that are unfamiliar and do not have clear cut answers is uncomfortable, often resulting in the person leaving the project or the company for a programming position elsewhere.

■ The characteristics of this new domain are not what they sought when they went into software engineering [8, 9].

■ Top performers see what happens within the computer as a logical reality and what happens outside of that domain to be illogical and somewhat annoying [9]. Managing will not always involve logical actions but requires "soft" skills which are rarely logical [10] and not often the strength of top performing software engineers in management roles [3].

Perhaps the worst part about this practice is that the company ends up converting a high performing software engineer into an often frustrated software project manager that would prefer to be writing code. In my experience, I have seen this happen at different firms with some software engineers requesting reclassification from manager back to software engineer. This practice of promoting the most accomplished software engineer into management is a double-edged sword in that the organization's costs go up, productivity goes down and members of the development team, being poorly managed leave the project or the company resulting in software development costs going up by as much as 60% [7]. If that sounds excessive, consider the following:

■ Whether the project replaces the departed developer from within the company or hires someone from outside the firm, the replacement will need some help becoming knowledgeable regarding the project, its plan, what has been achieved and where they fit in.

■ The replacement may not agree with the approach being taken by the project and therefore will be working to overcome cognitive dissonance which reduces their productivity [11].

■ Other people on the team will have to work with the replacement to get them "up to speed" which takes away from their productivity.

There are many other factors involved but suffice it to say the loss of a team member has many negative effects. While being a high Capability Maturity Model (CMM) organization or one with a mature, well defined, effective software development methodology in place providing a structure within which software is developed may reduce the impact of the loss of a software engineer, the fact remains, losing a portion of the software engineering team has a negative impact on the project(s) they are working on. Some software project managers view software engineers in much the same way they view hardware devices – plug compatible. In fact, software engineers are anything but. Like human kind, each is a unique individual with their own positive and negative aspects. This is why Brooks recommends naming specific people when developing a project plan [12].

4.3 Hard versus Soft Management Skills

A high turnover rate has a cause. The number one reason people leave a project team or a company is their project manager [7]. One might propose that those who decide to leave believe their technical skills and other abilities are far superior to those of their manager. But again, education comes into the picture. If the software engineer who wants to leave the company had been educated in the fundamentals of software project management, they would know that the software project manager's role is to perform five functions in collaboration with the software engineering team:

- ■ *Planning*: Putting together a list of tasks and subtasks necessary for the project to be successful including the names of those who will be responsible for completing them.
- ■ *Scheduling*: Time ordering the tasks and subtasks based on their interdependencies, customer requirements, final delivery date and other milestones together with the burdened costs associated with each as well as the final cost which includes profit.
- ■ *Controlling*: Monitoring progress using Earned Value Management, taking corrective action, as necessary to keep the project on track to achieve on time delivery.
- ■ *Staffing*: Identifying and acquiring the necessary software engineering skills needed to conduct the project.
- ■ *Motivating*: Taking action(s) that keep the project team engaged and self-motivated to do a good job and complete the project.

Notice that in the list of the five functional areas of software project management, each function is done continuously, switching between activities, as necessary, in order to maintain the project's momentum, revising the project plan or other documents as needed, conducting status meetings and so forth. Also, writing software is not one of the management functions nor is being the most technically astute member of the team. Some employment advertisements for software project managers indicate that the software project manager may be called upon to write software, as needed, throughout the project. This indicates that companies running such ads do not understand the nature of project management. It is a continuous activity, requiring a different mindset than writing software. As a result, having the software project manager interrupt her/his management activities to write code can jeopardize the project. Although the software project manager must have some knowledge of the technology being used, being the best programmer on the team has been shown not to be necessary for success [13]. In fact, the most successful high technology teams have three characteristics, none of which include having a super technology expert [13]:

- They must all have some knowledge of the technology being used
- They need to have experience working as part of a team
- They must have compatible personalities

If the team has incompatible personalities present, it is the software project manager who must deal with the resulting issues and make things work.

4.4 Hard versus Soft Skills

Software engineering began as a profession engaged in by some very technically astute people. In spite of its humble beginnings in the late 19th century using punched cards to speed the counting and analysis of the census of the United States of America, by the middle of the 20th century, the speed of the evolution into an advanced technology was inevitable. Those founders saw technology as the solution to a broad range of problems. When Boehm's book on Software Engineering Economics in 1984 [14] was published, it introduced the Constructive Cost Model (COCOMO) which ignored project management factors. Years later, Boehm admitted that he had assumed that a cadre of competent software project managers existed so that sector did not need to be addressed by COCOMO. In hindsight, we now know that software project managers are an important part of the fabric of Software Engineering. This was highlighted by the findings by IBM that 53% of failed projects failed due to, "Poor management" while 3% failed due to "Technical challenges." Perhaps due to the lack of well-defined criteria or simply old habits, companies still promote high performing software engineers into software project management roles. This has contributed to the software engineering industry exhibiting the highest rate of project failures in the world [15]. The usual factors causing these failures are listed by various authors. These include scope creep, spreading the development team over multiple projects, using the wrong lifecycle and many others. Rarely, if ever, is the lack of management education on the part of the software project manager mentioned. Instead, these and other excuses are put forward as a convenient firewall protecting some from responsibility for catastrophe. Seemingly unimportant factoids that a course in software project management would present include the results of a study of more than 700 projects (high technology to low technology) that showed that if a project is 15% complete and over budget and/or behind schedule its chances of finishing on time and/or within budget are nil [16]. It appears to be something of a "law" of project management. Education would also demonstrate that studies have shown that starting a software project with more people than you believe you need so you won't fall behind schedule also does not work [17]. While there are hundreds of other items of this nature, the fact remains that most software project managers have had no training in project management [3]. That fact alone explains many of the problems software projects experience. The reason(s) behind may include the fact that project management in general is a fluid, unpredictable and often experimental activity which is markedly different from writing and developing software. The "hard" skills such as exemplary programming talent an individual may have who is put into software project management actually work against them due to the differences between the two roles. Include areas software project managers have to concern themselves with such as the personnel issues involving evaluating people's performance, resolving interpersonal conflict, accommodating cultural differences and so forth and you have a technology expert put into a domain for which they may have no prior experience or training.

4.5 Changing the Current Situation

Enrollments in software engineering degree programs continue to increase as students begin to realize that employment in the software engineering profession continues to be stable with above average pay and benefits and expanding. While the curricula in most colleges and universities generally provide a solid grounding in the important aspects of software development, few require training in software project management. This places the graduate from one of these programs on a collision path between what industry needs and what the graduated student can provide. Due to their training in software engineering and their job performance, their management team will see them as a logical candidate for a software project management position. Mega corporations have their own management training programs but these are general management knowledge and focused on manager-employee interactions so as to avoid employee lawsuits for discrimination, harassment, bias and so forth. How training in software project management differs from "general management principles" training is that it takes into account the psychological profile that is unique to software engineers [8], the various options that the software project manager has in running a project and other matters that are unique to software engineering.

4.6 A Multicultural Management Practice

The practice of software engineering is not confined to people from English speaking, highly industrialized countries but has expanded to a broad range of countries who have provided educational opportunities to their youth who, having taken advantage of them, often leave their home country to work where the demand has dictated higher wages. The United States of America is an excellent example of a

Table 4.1   Where Foreign Born Software Engineers Working in the United States Come From [19]

Country	Number Employed in the U.S.	Relative %
India	23,195	40.8
China	7,680	13.5
Canada	3,427	6.0
Russia	3,349	5.9
South Korea	1,998	3.5
Vietnam	1,683	3.0
Philippines	1,194	2.1
Mexico	1,081	1.9
England	965	1.7
Germany	865	1.5

beneficiary of this migration (Table 4.1). But with these benefits come some often unexpected challenges. For example, for most of these software engineers, English is not their first language. This can result in miscommunication. [18]

With such a high probability of having to manage people from different backgrounds, value systems and cultures, software project managers need to be trained in the basics of managing conflict, how to motivate people from various cultures, multicultural negotiation and more. Note that little or none of this is technical in nature with interpersonal communication skills being a greater asset than technical skill. Again, our de facto policy of putting the most technically skilled individuals into management positions helps explain why so many software projects fail due to, "poor management."

4.7 Managing Multiple Generations

The 21st century may mark the first time in industrialized countries that multiple generations will be working together. Often, the software project manager will be managing a team of software engineers many of whom are older than the manager. It turns out, depending on their age, different team members may have significantly different value systems from each other as well as their manager. This leads to unexpected resistance to decisions and policies by the software project manager. The response of the software project manager is often frustration and puzzlement as to why what seems to be a "no brainer" causes such consternation within the team. Being trained as to what the cause of this resistance is can go a long way to improving communication between the software project manager and the team [20].

4.8 What Is a Successful Software Engineering Team?

Since the start of the competition to put a man on the moon and safely bringing him back to earth, teams of people have been created to solve challenging technical problems. Lacking information on the success or failure of various schemes for team composition, senior managers reasoned that teams that would be the most proficient at solving these problems would be composed of the highest performing engineers. We now know that they were wrong due to something called, "The Apollo Syndrome" [21]. It turns out that the highest performers tend to see themselves as more skilled than their colleagues and as a result are difficult to work with. This makes teaming to solve problems unlikely to be successful. More recently, work has been done to study successful teams identifying the characteristics they share. This gives us insight into what works based on evidence [13]. While the results are based on overwhelming evidence, they are nonetheless surprising. What the researchers found was that successful high technology teams all exhibited three characteristics:

1. They all had some knowledge of the technology involved but none of them could be considered an expert in the field. The significance of this may be that the team members are all discovering nuances of the technology they are dealing with rather than having an expert who might dictate the path to progress.
2. They had experience working as part of a team. This means they had experience with the give and take that occurs in a technical team as well as the compromises that are made to achieve success. They have experienced the mutual support so vital to a team's success.
3. They had compatible personalities. Using the Myers–Briggs Type Indicator (MBTI) [22], they found that compatible personalities were also vital to success. In this case, incompatible personalities will result in friction within the team wasting vital flow time and energy rather than forging progress.

The three factors listed above are like the legs of a stool. Remove one and the stool falls over. Similarly, lacking one of these characteristics jeopardizes the success of the team. For example, having team members who cannot work to mutually support each other due to egotism or incompatible personalities means the team is unlikely to achieve success.

4.9 Conclusion

At the present time, software engineering can be considered a relatively immature profession as evidenced by its seemingly continual search for the next new method to solve the "software engineering" problem (i.e., projects consistently being delivered late, over budget and not meeting all requirements). Part of its problems stem from a pronounced lack of recognition of the importance of the software project manager to a successful software project. The data is there if only we alter our perception to accept it and take meaningful action.

References

1. LaPlante, P., "Antipatterns: Identification, Refactoring and Management," Auerbach Publications, New York, N.Y., 2005.
2. Gulla, J., "Seven Reasons Why IT Projects Fail," IBM Systems, 01 February 2012.
3. Katz, R., "Motivating Technical Professionals Today," IEEE Engineering Management Review, Vol. 41, No. 1, pp. 28–38, 2013.
4. Townsend, R., "Up the Organization: How to Stop the Corporation from Stifling People and Strangling Profits," Jossey-Bass, 1st edition, 06 January 1970.
5, Gardner, H., Gino, F., Staats, B.R., "Dynamically Integrating Knowledge in Teams: Transforming Resources into Performance," Working Paper 11-009, Harvard Business School, 07 September 2011.
6. Staats, B.R., Gino, F., Pisano, G.P., "Varied Experience, Team Familiarity, and Learning: The Mediating Role of Psychological Safety," Working Paper 10-016, Harvard Business School, 2010.
7. Cone, E., "Managing that Churning Sensation," Information Week, No. 680, pp. 50–67, May 1998.
8. Couger, D.J. and Zawacki, R.A., "Motivating and Managing Computer Personnel," Wiley-Interscience, New York, N.Y., 1980.
9. Turkle, S., "The Second Self: Computers and the Human Spirit," MIT Press, Cambridge, MA 2005.
10. Sukhoo, A., Barnard, A., Eloff, M.M., Van der poll, J.A., "Accommodating Soft Skills in Software Project Management," Issues in Informing Science and Information Technology, January 2005.
11. Weinberg, G., "The Psychology of Computer Programming," Weinberg and Weinberg, 2011.
12. Brooks, F., "The Mythical Man-Month – Essays on Software Engineering," Addison Wesley Professional Series, 2nd edition, 1995.
13. Chen, J. and Lin, L., "Modeling Team Member Characteristics for the Formation of a Multifunctional Team in Concurrent Engineering," IEEE Transactions on Engineering Management, Vol. 15, No. 2, pp. 111–124, 2004.
14. Boehm, B., "Software Engineering Economics," Prentice-Hall, Englewood Cliffs, N.J., 1984, pp. 486–487.
15. Dorsey, P., "Top 10 reasons why systems projects fail," Dulcian.com, 26 Apr 2005.
16. Fleming, Q.W. and Koppelman, J.M., "Earned Value Project Management," Project Management Institute, Newtown Square, Pennsylvania, 4th edition, 2010.

17. Staats, B.R., Milkman, K.L. and Fox, C.R., "The Team Sizing Fallacy: Underestimating The Declining Efficiency of Larger Teams," Forthcoming article in Organizational Behavior and Human Decision Processes, Vol. 118, 2012.
18. Patel, D., Lawson-Johnson, C. and Patel, S., "The Effect of Cultural Differences on Software Development," 5th International Conference on Computing and ICT Research (ICCIR'09), Kampal, Uganda, pp. 250–262, August, 2009.
19. Balk, G., "More than half of Seattle's Software Developers were born outside U.S.," Seattle Times, 19 January 2018.
20. Knight, R., "Managing People from 5 Generations," Harvard Business Review, September, 25, 2014.
21. Belbin, M., "Management Teams – Why They Succeed or Fail," Butterworth-Heinemann, 23 April Waltham, MA. 1996.
22. Myers, I.B. and Myers, P.B., "Gifts Differing," CPP, 2nd edition, Davies-Black, Palo Alto, CA., 1995.

Chapter 5

Inter-Parameter Dependencies in Real-World Web APIs: The IDEA Dataset

Alberto Martin-Lopez[a], Sergio Segura[a], and Antonio Ruiz-Cortés[a]

[a]SCORE Lab, I3US Institute, University of Seville, Spain

Contents

5.1 Background and Summary 101
5.2 Dataset Specification..103
5.3 Value of the Data 104
5.4 Data Description..104
5.5 Experimental Design, Materials, and Methods 105
5.6 Competing Interests..105
Acknowledgment 105
References 105

5.1 Background and Summary

Web services are heavily used nowadays for integrating heterogeneous software systems. In this context, many contributions have been made to automate several processes such as service design, discovery, composition and testing. However, such approaches still suffer to a large extent from one fundamental problem: the management of inter-parameter dependencies. An inter-parameter dependency imposes a constraint on how two or more input parameters can be combined to form valid calls to the service. For example, in the YouTube API, when searching for videos in high definition (parameter videoDefinition), the parameter type must be set to "video", otherwise an HTTP 400 status code ("bad request") is returned in the response. Unfortunately, current API specification languages such as the OpenAPI Specification (OAS) [3] or the RESTful API Modeling Language (RAML) [5] provide no support for the formal description of such dependencies. The interest of industry in having support for these is reflected in an open feature request in OAS entitled "Support interdependencies between query parameters" [2], created in January 2015 with the message shown below. At the time of this writing, the request has received over 280 votes and 55 comments from 33 participants:

> *It would be great to be able to specify interdependencies between query parameters. In my app, some query parameters become 'required' only when some other query parameter is present. And when conditionally required parameters are missing when the conditions are met, the API fails. Of course I can have the API reply back that some required parameter is missing, but it would be great to have that built into Swagger.*

This feature request has fostered an interesting discussion where the participants have proposed different ways of extending OAS to support dependencies among input parameters. However, each approach aims to address a particular type of dependency and thus show a very limited scope. Addressing the problem of modeling and validating inter-parameter dependencies in web APIs should necessarily start by understanding how dependencies emerge in practice. To address this problem, in a previous paper [9], we performed a thorough study on the presence of inter-parameter dependencies in industrial web APIs. Our study is based on an exhaustive review of 2,557 operations from 40 APIs selected among ten different application domains. As a result, we managed to generate a dataset containing all inter-parameter dependencies found (633 in total) classified into seven general types.

This chapter presents the Inter-parameter DEpendencies in web Apis (IDEA) dataset, generated as a part of our previous survey on the presence of inter-parameter dependencies in web APIs [9]. Among other results, we found that inter-parameter dependencies are extremely common and pervasive: they appear in four out of every five APIs across all application domains and types of operations. Researchers and practitioners can leverage the IDEA dataset for multiple purposes. For instance, proposals for modeling and validating dependencies may benefit from the results obtained in our survey. Furthermore, new approaches for web services testing and discovery may involve the management of inter-parameter dependencies to achieve a higher degree of automation.

5.2 Dataset Specification

Subject	Web Application Programming Interfaces (APIs).
Specific Subject Area	Specification of web APIs.
Type of Data	Spreadsheets.
How Data Were Acquired	Survey comprising 2,557 operations from 40 real-world web APIs. The online documentation of each API was carefully read and analyzed by at least two people to reduce misunderstanding. *Instruments used*: Standard PC with Internet access for reviewing the documentation of the APIs.
Data Format	Analyzed.
Parameters for Data Collection	Only APIs complying with level 1 of the Richardson Maturity Model (i.e., "REST-like APIs") [6, 11] were considered. A total of 2,557 operations (including read, write, update and delete operations) from 40 APIs were selected from the ProgrammableWeb API repository [4], including the ten most popular APIs and the three most popular APIs from the ten most popular categories.
Description of Data Collection	Inter-parameter dependencies were identified in two steps. First, the online documentation of each API was carefully read and analyzed, recording all dependencies found. This was done by at least two authors per API, in order to reduce mistakes. In a second step, the shape of all dependencies was studied, and they were classified into seven general patterns.
Data Source Location	Institution: Applied Software Engineering (ISA) research group. University of Seville. City: Seville. Country: Spain.
Data Accessibility	Repository name: IDEA Dataset: Inter-parameter DEpendencies in web Apis. Direct URL to data: https://bit.ly/35aqBBS
Related Research Article	A. Martin-Lopez, S. Segura, A. Ruiz-Cortés. A Catalogue of Inter-Parameter Dependencies in RESTful Web APIs. International Conference on Service-Oriented Computing (2019). https://doi.org/10.1007/978-3-030-33702-5_31

5.3 Value of the Data

■ The IDEA dataset is key for the correct understanding of how inter-parameter dependencies are used in practice. It contains samples of all possible types of dependencies for which practitioners are asking support in OAS (and more). It is also machine-readable.

■ Both researchers and practitioners can benefit from these data. Researchers on service-oriented architectures may leverage the dataset to automate the analysis of dependencies in web APIs. Practitioners may be interested to integrate the specification and analysis of dependencies in current standards such as OAS [2, 3] and RAML [5].

■ The IDEA dataset offers ample research opportunities in the field of service-oriented computing. Proposals for the specification and automated analysis of dependencies can be tackled. The fact that it is machine-readable that can help automate processes such as web service discovery and composition.

■ The IDEA dataset has already proved useful in some contexts. For example, we found that the misspecification of dependencies is a common problem in web APIs [7, 8]. Furthermore, the coverage of valid and invalid combinations of input parameters could be considered a new test coverage criterion for web APIs [10].

5.4 Data Description

The dataset is available online as supplementary material of our survey [1]. It consists of a Google spreadsheet containing 67 tabs, described here:

■ The first tab ("Home") includes information regarding the most relevant tabs in the spreadsheet.

■ The second tab ("Subjects APIs full analysis") contains information about the 40 APIs analyzed, including the name, category, URL to the documentation and number and percentage of operations with or without dependencies classified by type (i.e., their CRUD semantic), among others.

■ The third tab ("Selection process") provides a summary of the selection process followed for filtering the APIs from ProgrammableWeb [4].

■ Tabs 4–12 include statistics and charts regarding the size (number of operations), category and number of dependencies of the subject APIs.

■ Tabs 13–27 include statistics and charts regarding the type and shape of all the dependencies found, such as the number and type of parameters involved (e.g., query and header parameters).

■ Tabs 28–67 include details about all inter-parameter dependencies found in every API under study (40 tabs in total, one per API).

5.5 Experimental Design, Materials, and Methods

The subject APIs were selected from the popular API repository ProgrammableWeb [4]. More specifically, we selected the top ten most popular APIs and the three top-ranked APIs from the ten most popular categories. We selected APIs reaching level 1 or higher in the Richardson Maturity Model [6, 11], which ensured a minimal adherence to the REST architectural style, the current de facto standard for web API design. Overall, we selected 40 APIs distributed among ten different categories such as communication, social and mapping, including web APIs with millions of users worldwide such as Google Maps and PayPal. Regarding the size, the majority of reviewed APIs (75%) provide between 1 and 50 operations, with the largest APIs having up to 305 (DocuSign eSignature) and 492 (GitHub) operations.

Dependencies were identified in two steps. First, we recorded all the inter-parameter dependencies found in the documentation of the subject APIs. It is worth mentioning that every dependency can be represented in multiple ways, e.g., in conjunctive normal form. At this point, we strove to represent them as they were described in the documentation of the API. This allowed us, e.g., to record the *arity* of each dependency, i.e., number of parameters involved in each constraint. In a second step, we studied the shape of all the dependencies and managed to group them into seven general dependency types. The documentation collected from each API was reviewed by at least two different authors to reduce misunderstanding or missing information.

5.6 Competing Interests

The authors declare that they have no known competing financial interests or personal relationships which have, or could be perceived to have, influenced the work reported in this article.

Acknowledgment

This work has been partially supported by the European Commission (FEDER) and the Spanish Government under projects HORATIO (RTI2018-101204-B-C21), APOLO (US-1264651), EKIPMENT-PLUS (P18-FR-2895), and the FPU scholarship program, granted by the Spanish Ministry of Education and Vocational Training (FPU17/04077).

References

1. A. Martin-Lopez, S. Segura, A. Ruiz-Cortés, IDEA Dataset: Inter-parameter DEpendencies in web Apis (Version v1) [Data set], Zenodo (2021). http://doi.org/10.5281/zenodo.4495136
2. Feature request in OAS: "Support interdependencies between query parameters". https://github.com/OAI/OpenAPI-Specification/issues/256, 2015 (accessed April 2020).

3. OpenAPI Specification. https://www.openapis.org/, 2020 (accessed April 2020).

4. ProgrammableWeb API repository. https://www.programmableweb.com/, 2020 (accessed April 2020).

5. RAML: RESTful API Modeling Language. https://raml.org/ (accessed April 2020).

6. Richardson Maturity Model. https://martinfowler.com/articles/richardsonMaturity-Model.html, 2010 (accessed April 2020).

7. A. Martin-Lopez, AI-Driven Web API Testing, International Conference on Software Engineering: Companion Proceedings (2020), 202–205. https://doi.org/10.1145/3377812.3381388

8. A. Martin-Lopez, Automated Analysis of Inter-Parameter Dependencies in Web APIs, International Conference on Software Engineering: Companion Proceedings (2020). 140–142. https://doi.org/10.1145/3377812.3382173

9. A. Martin-Lopez, S. Segura, A. Ruiz-Cortés, A Catalogue of Inter-Parameter Dependencies in RESTful Web APIs, International Conference on Service-Oriented Computing (2019), 399–414. https://doi.org/10.1007/978-3-030-33702-5_31

10. A. Martin-Lopez, S. Segura, A. Ruiz-Cortés, Test Coverage Criteria for RESTful Web APIs, International Workshop on Automating TEST Case Design, Selection, and Evaluation (2019), 15–21. https://doi.org/10.1145/3340433.3342822

11. L. Richardson, S. Ruby, M. Amundsen, RESTful Web APIs, 1st edition, O'Reilly Media, Inc., 2013.

Chapter 6

Evaluating Testing Techniques in Highly-Configurable Systems: The Drupal Dataset

Ana B. Sánchez[a], Sergio Segura[b],
and Antonio Ruiz-Cortés[b]

[a]*University of Seville, Seville, Spain*
[b]*SCORE Lab, I3US Institute, University of Seville, Spain*

Contents

6.1   Background and Summary 108
6.2   Dataset Specification 109
6.3   Value of the Data 111
6.4   Data Description 111
6.5   Experimental Design, Materials, and Methods 111
6.6   Conclusion 113
Acknowledgment 113
References 113

6.1 Background and Summary

Highly-Configurable Systems (HCSs) determine the ability of software applications to be extended, customized or configured [1]. Operating systems such as Linux [2] or development tools such as Eclipse [3] have been reported as examples of HCSs. Another prominent example of HCS is Software Product Lines (SPL) [4]. SPL engineering focuses on the development of families of related products through the systematic management of variability. For that purpose, *feature models* are typically used as the de facto standard for configurability modeling in terms of functional features and constraints among them [5]. The number of configurations in these models is potentially huge. This makes testing HCSs a challenge task. To address this problem, researchers have proposed numerous techniques to reduce the cost of testing in the presence of variability [1, 3, 6–9]. To evaluate these techniques, unrealistic experiments are usually carried out by researchers that employ synthetic feature models and data that introduce threats to validity and question their conclusions. This is due to the lack of real-world data available about HCSs that share code, test cases, detailed fault report or even a detailed documentation that enables the reproducibility of experiments in realistic environments [2].

In order to search for real-world HCSs with available code we followed the steps of previous authors and looked into the open source community. Particularly, we found the popular open-source Drupal framework, a highly modular web content management written in PHP [10, 11]. Drupal has more than 30,000 modules that can be composed to form valid configurations of the system. Drupal provides detailed fault reports including fault description, fault severity, type, status and so on. The high number of the Drupal community members together with its extensive documentation have also been strengths to choose this framework, currently maintained and developed by a community of more than 630,000 users and developers. Drupal can be used to build a variety of web sites including Internet portals, e-commerce applications and online newspapers. Drupal is composed of a set of modules. A module is a collection of functions that provide certain functionality to the system. According to the Drupal documentation, each module that is installed and enabled adds a new feature to the framework [11]. Thus, we propose modeling Drupal modules as features of the feature model.

In this chapter, we present the Drupal dataset, a publicly available resource of valuable testing data about the Drupal framework and its modules. In particular, we provide the following information:

1. *The Drupal feature model:* We model some of the main Drupal modules to features and represent the framework configurability using a feature model. The resulting model has 48 features, 21 cross-tree constraints and represents 2.09E9 different Drupal configurations.

2. *Non-functional Drupal data:* We report on extensive non-functional data extracted from the Drupal Git repository. For each feature under study, we report its size, number of changes (during 2 years), cyclomatic complexity, number of test cases, number of test assertions, number of developers and

number of reported installations. The non-functional data are modeled as feature attributes in the feature model.

3. *History of Drupal faults:* We present the number of faults reported on the Drupal features under study during a period of 2 years, extracted from the Drupal issue tracking system. Faults are classified according to their severity and the feature(s) that trigger it. Among other results, we identified 3,392 faults in Drupal v7.23, 160 of them caused by the interaction of up to four different features. We replicated the study of faults in two consecutive Drupal versions, v7.22 and v7.23, to enable fault history test validations.

As far as we know, the Drupal dataset has been used by numerous research articles to evaluate their HCS testing techniques. Proof of this are the 150 citations received since its first research publications [12–14]. Among others researches, Hierons et al. proposed a new testing technique to solve the many-objective optimization problem that was evaluated with the Drupal dataset [15]. Previously, the same authors presented a proposal for obtaining the optimal selection of products from feature models using many-objective evolutionary optimization. The feasibility of this approach was assessed with data that included the Drupal dataset [16]. Fischer et al. performed an empirical assessment of similarity for testing software product lines and the Drupal dataset was key in their evaluation [9].

6.2 Dataset Specification

Subject	Testing of highly-configurable systems
Specific Subject Area	Automated test case selection, prioritization and minimization of highly-configurable systems.
Type of Data	Table
	Model
	Figure
How Data Were Acquired	Repository and bug tracking mining and literature review Software, console commands and manual data review
Data Format	Link to the Drupal DataSet and experiments: https://github.com/belene/DrupalDataset Contains: Drupal feature model in SXFM format (XML format) Drupal feature model in FaMa format (XML format) Drupal feature data (CSV format) Drupal feature faults (CSV format) Analyzed Filtered

Parameters for Data Collection	We designed the Drupal feature module using 48 modules and 21 dependencies identified among the modules. To analyze the effectiveness of non-functional data as bug predictors, we collected the commits made in modules and the faults recorded in two different versions of Drupal and in the period of 2 years, obtaining a total of 557 changes and 3,301 faults for Drupal v7.22.
Description of Data Collection	We followed a systematic approach and mapped each Drupal module to a feature. We also used the dependencies defined in the information file of each Drupal module to model cross-tree constraints. We filtered the faults by feature name, framework version and dates. Then, the search was refined to eliminate the faults not accepted by the Drupal community. Next, we manually checked the bug reports of each candidate integration fault and discarded those not correctly identified. To obtain the number of changes made in each feature, we tracked the commits to the Drupal Git repository by using console commands. The rest of the data was obtained through the review of the official Drupal documentation and websites.
Data Source Location	Institution: University of Seville City: Seville Country: Spain
Data Accessibility	The data are publicly available in a repository. But part of the data and their explanations are included in the article. Repository name: The Drupal Dataset Direct URL to data: https://github.com/belene/DrupalDataset
Related Research Article	Sánchez, Ana B., Segura, Sergio, Parejo, José A., Ruiz-Cortés, Antonio. Variability testing in the wild: The drupal case study. Software & Systems Modeling. 1 (2017) 173–194. https://doi.org/10.1007/s10270-015-0459-z Sánchez, Ana B., Segura, Sergio, Ruiz-Cortés, Antonio. The drupal framework: A case study to evaluate variability testing techniques. Proceedings of the Eighth International Workshop on Variability Modelling of Software-Intensive Systems. https://doi.org/10.1145/2556624.2556638

6.3 Value of the Data

- Drupal dataset enables the evaluation of testing techniques for HCSs with real-world data from the community open-source.
- Both, researchers and practitioners can benefit from these data. This dataset provides variability researchers and practitioners with helpful information about the distribution of faults and test cases in a real HCS. Also, it is a valuable asset to evaluate HCS testing techniques in realistic settings rather than using random variability models and simulated faults.
- The data collected in the Drupal dataset can be used as good indicators of the fault propensity of a software application by using the history of faults and changes in different versions of Drupal.
- The Drupal dataset can be useful in any analysis work of feature models.

6.4 Data Description

The Drupal dataset is publicly available in a Github repository: https://github.com/belene/DrupalDataset

- ***DRUPALv4SXFM.xml:*** Drupal feature model in SXFM format. The model comprises 48 features and 21 constraints. It can be found in the folder FMs of the DrupalDataset repository.
- ***DRUPALv4FAMA.xml:*** Drupal feature model in FaMa format. The model comprises 48 features and 21 constraints. It can be found in the folder FMs of the DrupalDataset repository.
- ***DrupalFeaturesData.csv:*** A CSV File containing the Drupal modules considered in the dataset and, for each module, the size, the code cyclomatic complexity, the number of test cases and assertions, the number of reported installations, the number of commits made in Drupal versions v7.22 and v7.23, and the number of faults recorded in Drupal v7.22 and v7.23.
- ***DrupalFeatureFaults.csv:*** A CSV file containing for each Drupal module the number of collected faults classified by type (single or integration faults), severity (minor, normal, major and critical) and by Drupal version (v7.22 and v7.23).

6.5 Experimental Design, Materials, and Methods

Drupal feature model: According to the Drupal documentation, each module that is installed and enabled in the system adds a new feature to the framework [11]. Thus, we followed a systematic approach and proposed modeling Drupal modules as features of the feature model. Also, when a module is installed, new subfeatures can be enabled adding extra functionality to the module. These features are considered as children features of the module that contains them. The Drupal core modules that must be always enabled are represented as mandatory relations in the feature model. On the other side, all the modules that can be optionally installed and enabled in the system are modeled as optional features in the model. In addition to this, Drupal modules can have dependencies with other modules, i.e., modules that must be installed and enabled for another module to work properly. These dependencies are modeled as cross-tree constraints in the form of requires in the feature model.

Non-functional data: Drupal dataset also reports a number of non-functional attributes of the features selected for the Drupal feature models, namely:

- **Feature size:** This provides a rough idea of the complexity of each feature and its fault propensity. The size of a feature was calculated in terms of the number of lines of code (LoC).
- **Cyclomatic Complexity (CC):** This metric reflects the total number of independent logic paths used in a program and provides a quantitative measure of its complexity. We used the open-source tool *phploc* to compute the CC of the source code associated with each Drupal feature. Roughly speaking, the tool calculates the number of control flow statements (e.g., "if", "while") per lines of code.
- **Number of tests:** We provide the total number of test cases and test assertions of each Drupal feature obtained from the output of the *SimpleTest* module.
- **Number of reported installations:** This depicts the number of times that a Drupal feature has been installed as reported by Drupal users. This data was extracted from the Drupal website [10] and could be used as an indicator of the popularity or impact of a feature.
- **Number of developers:** We collected the number of developers involved in the development of each Drupal feature. This could give us information about the scale and relevance of the feature as well as its propensity to faults related to the number of people working on it. This information was obtained from the website of each Drupal module as the number of committers involved [10].
- **Number of changes:** Changes in the code are likely to introduce faults. Thus, the number of changes in a feature may be a good indicator of its error proneness and could help us to predict faults in the future. To obtain the number of changes made in each feature, we tracked the commits to the Drupal Git repository.

Faults in Drupal. The Drupal dataset collects the number of faults reported in the Drupal features. The information was obtained from the issue tracking systems of Drupal and related modules. We used the web-based search tool of the issue systems to filter the bug reports by severity, status, date, feature name and Drupal version. The search was narrowed by collecting the bugs reported in a period of 2 years and in two consecutive Drupal versions, v7.22 and v7.23, to achieve a better understanding of the evolution of a real system and to enable test validations based on fault history. Then, the search was refined to eliminate the faults not accepted by the Drupal community, those classified as duplicated bugs, non-reproducible bugs and bugs working as designed. Additionally, we identified and classified the faults into single (those caused by a Drupal model) and integration faults (those caused by the interaction of several modules). To mitigate possible misidentification of faults, we manually checked each candidate integration fault to discard those that did not correspond.

For the sake of validation, the work was discussed with two Drupal core team members who approved the followed approach. We also may mention that the main author of the article has more than 2 years of experience in industry as a Drupal developer.

6.6 Conclusion

The Drupal dataset presented in this paper consists of a set of real-world data extracted from the well-known Drupal framework. This dataset is a valuable asset to evaluate testing techniques of HCSs in realistic settings. To the best of our knowledge, the Drupal dataset has been used by numerous researchers to assess their HCS testing techniques as it is evidenced by the 150 citations received since its first publication [12–14].

Acknowledgment

This work has been partially supported by the European Commission (FEDER), the Spanish Government under the project HORATIO (RTI2018-101204-B-C21) and the Andalusian R&D&I programs APOLO (US-1264651) and EKIPMENT (PR18-FR-2895).

References

1. Cohen, M. B., Dwyer, M. B., Shi, J. Constructing interaction test suites for highly-configurable systems in the presence of constraints: A greedy approach. Transactions on Software Engineering. 5 (2008) 633–650. https://doi.org/10.1109/TSE.2008.50.
2. She, S., Lotufo, R., Berger, T., Wasowski, A., Czarnecki, K. The variability model of the linux kernel. International Workshop on Variability Modelling of Software-Intensive Systems. 2010.
3. Johansen, M. F., Haugen, O., Fleurey, F., Eldegard, A. G., Syversen, T. Generating better partial covering arrays by modeling weights on sub-product lines. International Conference on Model Driven Engineering Languages and Systems. 2012. pp. 269–284. https://doi.org/10.1007/978-3-642-33666-9_18.

4. Svahnberg, M., van Gurp, L., Bosch, J. A taxonomy of variability realization techniques: Research articles. Softw. Pract. Exp. 35 (2005) 705–754. https://dl.acm.org/doi/10.5555/1070904.1070905.

5. Kang, K., Cohen, S., Hess, J., Novak, W., Peterson, S. Feature-Oriented Domain Analysis (FODA) Feasibility Study. Software Engineering Institute. 1990.

6. Yoo, S., Harman, M. Regression testing minimisation, selection and prioritisation: A survey. Software Testing, Verification and Reliability. (2012) 67–120. https://doi.org/10.1002/stv.430.

7. Henard, C., Papadakis, M., Perrouin, G., Klein, J., Heymans, P., Le Traon, Y. Bypassing the combinatorial explosion: Using similarity to generate and prioritize t-wise test configurations for software product lines. IEEE Trans. Softw. Eng. 7 (2014) 650–670.

8. Sánchez, A. B., Segura, S., Ruiz-Cortés, A. A comparison of test case prioritization criteria for software product lines. IEEE International Conference on Software Testing, Verification, and Validation. 2014, pp. 41–50. https://doi.org/10.1109/ICST.2014.15.

9. Fischer, S., Lopez-Herrejon, R. E., Ramler, R., Egyed, A. A preliminary empirical assessment of similarity for combinatorial iteraction testing of software product lines. IEEE/ACM 9th International Workshop on Search-Based Software Testing (SBST), 2016, pp. 15–18.

10. Buytaert, D. Drupal framework. http://www.drupal.org. Accessed April 2020.

11. Tomlinson, T., VanDyk, J. K. Pro Drupal 7 Development. Tomlinson, Todd (et al.). 3rd edition. 2010.

12. Sánchez, Ana B., Segura, Sergio, Ruiz-Cortés, Antonio. The drupal framework: A case study to evaluate variability testing techniques. Proceedings of the Eighth International Workshop on Variability Modelling of Software-Intensive Systems. 2014, pp. 1–8. https://doi.org/10.1145/2556624.2556638.

13. Sánchez, Ana B., Segura, Sergio, Parejo, José A., Ruiz-Cortés, Antonio. Variability testing in the wild: The drupal case study. Software & Systems Modeling. 1 (2017) 173–194. https://doi.org/10.1007/s10270-015-0459-z.

14. Parejo, José A., Sánchez, Ana B., Segura, Sergio, Ruiz-Cortés, Antonio, Lopez-Herrejon, Roberto E., Egyed, Alexander. Multi-objective test case prioritization in highly configurable systems: A case study. Journal of Systems and Software. (2016) 287–310. https://doi.org/10.1016/j.jss.2016.09.045.

15. Hierons, Robert M., Li, Miqing, Liu, Xiaohui, Parejo, Jose Antonio, Segura, Sergio, Yao, Xin. Many-objective test suite generation for software product lines. ACM Trans. Softw. Eng. Methodol. 1 (2020) 46 pages. https://doi.org/10.1145/3361146.

16. Hierons, Robert M., Li, Miqing, Liu, Xiaohui, Segura, Sergio, Zheng, Wei. SIP: Optimal product selection from feature models using many-objective evolutionary optimization. ACM Trans. Softw. Eng. Methodol. 17 (2016) 39 pages. https://doi.org/10.1145/2897760.

Chapter 7

A Family of Experiments to Evaluate the Effects of Mindfulness on Software Engineering Students: The MetaMind Dataset

Beatriz Bernárdez[a], Margarita Cruz[a],
Amador Durán[b], José A. Parejo[b],
and Antonio Ruiz-Cortés[b]

[a]*I3US Institute, Universidad de Sevilla, Spain*
[b]*SCORE Lab, I3US Institute, Universidad de Sevilla, Spain*

Contents

7.1   Background and Summary 116
7.2   Dataset Specification 118
7.3   Value of the Data 119
7.4   Data Description 120
7.5   Experimental Design, Materials, and Methods 121
7.6   About Benefits of Mindfulness in Software Engineering Activities 122
7.7   Competing Interest 122
Acknowledgment 122
Notes 123
References 123

7.1 Background and Summary

Mindfulness is a meditation technique whose main goal involves maintaining a calm mind and training attention by focusing only on one single thing (the support) at a time, this support is usually the practitioner's breathing. During a session of mindfulness, practitioners remain in a quiet location, program an alarm, and focus their attention on the support. When other thoughts come to mind, they are simply acknowledged, set aside, and the practitioners return their attention to their breathing. Improved attention, concentration, and mental clarity have been shown to be present among the benefits of mindfulness [1].

The reported benefits of its continued practice can be of interest for Software Engineering students and practitioners, especially in tasks such as conceptual modeling, in which concentration and clearness of mind are crucial. In order to ascertain whether the practice of mindfulness improves the students' performance in conceptual modeling tasks, a family of three controlled experiments has been carried out in the Software Engineering faculty of the University of Seville, Spain.

The idea of the original experiment (Mind#1) is to evaluate whether second-year Software Engineering students improved their performance in conceptual modeling subsequent to the daily practice of mindfulness over a period of 4 weeks, by comparing their results with those of a control group receiving a placebo treatment.

In order to achieve this goal, in Mind#1, the students were split into two groups: experimental and control. The experimental group was involved in a mindfulness workshop for 4 weeks. All the students developed two conceptual modeling exercises[1], one before starting the mindfulness workshop and a second when the workshop had finished. The students' evolution was analyzed in terms of effectiveness (the rate of correctly identified model elements) and efficiency (the number of correctly identified model elements per unit of time). The results showed that students who practiced mindfulness were statistically significantly more efficient than those of the control group. With respect to effectiveness, several improvements were observed but were not statistically significant [2].

Mind#1 was executed during the 2013–2014 academic year. Two internal replications (Mind#2 and Mind#3) were carried out in successive academic years with a double purpose: to confirm the results of Mind#1, and to overcome the limitations of the original experimental design [3].

One of the main limitations was related to the conceptual modeling exercises: after analyzing Mind#2, it was detected that certain unnoticed characteristics of such exercises could influence the outcome of the experiments. Therefore, in order to study such impact, it was decided to change the order of the conceptual modeling exercises in Mind#3. Hence, the students carried out the task involving

EoDP before the mindfulness workshop and then carried out the task involving ERASMUS after said workshop.

Once Mind#3 had finished, in order to analyze the whole family of experiments, a meta-analysis was performed, that is, a pooled data meta-analysis, in which the data of three studies are analyzed as if they had originated from a single study, and an aggregated data meta-analysis, which usually focuses on the analysis of the effect size and its variation among the studies.

The results of such meta-analysis reported in [4] revealed that both the effectiveness and efficiency of the students who practiced mindfulness were statistically significantly better than those of the control group.

The present chapter reports the dataset of the family of experiments, called MetaMind, employed to perform such a meta-analysis. The MetaMind dataset was generated as the aggregation of the three datasets of individual experiments (Mind#1, Mind#2, and Mind#3), but included certain new information deemed useful for the formation of a more conclusive meta-analysis.

Figure 7.1 shows the process of collecting, transforming, and analyzing the MetaMind dataset. Initially, for Mind#1, Mind#2, and Mind#3, the data collection was manually recorded using spreadsheets. After that, a filtered csv file was subsequently generated from these three spreadsheets. All the csv files were then automatically processed for their integration into a single file containing the simplified dataset of the whole family of experiments. During this process, information regarding the academic year (2014, 2015, or 2016) and the order in which the exercises were carried out in each individual experiment were added to the dataset. The csv file was used for the analysis in R-Studio. Furthermore, in order to enable most of the analyses to also be executed in SPSS, the data was imported to SPSS and saved as a file with extension ".sav", that is, the data file type in SPSS.

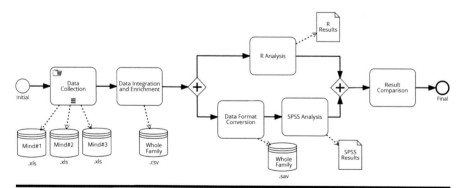

Figure 7.1 Process of collection and transformation of the MetaMind dataset reported in this chapter.

7.2 Dataset Specification

Subject	Effect of mindfulness on students' performance in conceptual modeling task. A Family of experiments.
Specific Subject Area	Students' effectiveness and efficiency in a conceptual modeling task, for both pooled and aggregated meta-analyses of the family of experiments.
Type of Data	csv files and sav files (SPSS).
How Data were Acquired	The effectiveness and efficiency scores were acquired manually, by means of a careful reading of the UML class diagrams developed by the students as their solution to the conceptual modeling exercises. After evaluating each student's solution, each score was labeled either as Pre-Treatment when the solution was obtained before the treatment, or Post-Treatment otherwise.
Data Format	Analyzed.
Parameters for Data Collection	To participate in the sample, students must fulfill the following selection criteria: (i) to attend at least 75% of the mindfulness sessions; (ii) to perform both ERASMUS and EoDP tasks; and (iii) to employ the correct notation in the conceptual modeling exercises, that is, UML class diagrams. The number of subjects that fulfilled the selection criteria was 36 in Mind#1, 53 in Mind#2, and 45 in Mind#3. As a consequence, there were 130 subjects in the sample for meta-analysis. The experimenters have accorded a *reference model* for each conceptual modeling exercise (ERASMUS and EoDP). Such models have been employed to evaluate the student's solution by assigning it a score.
Description of Data Collection	Data collection was carried out through the evaluation of the conceptual modeling solutions presented by the students before and after the mindfulness sessions. Model elements in a student's solution were considered as correctly identified if there existed identical or semantically equivalent counterparts in the corresponding *reference model*. For the effectiveness response variable, the score is the number of model elements (UML classes, attributes, and associations) correctly identified over the number of elements in the *reference model*. For the efficiency response variable, the score is the number of model elements correctly identified over the time spent solving the exercise.

Files	Mind#1: data2014.csv, Mind#2: data2015.csv, Mind#3: data2016.csv Meta-analysis (R-Studio): simplifieddata.csv Meta-analysis (SPSS): fulldata.sav and fulldataforancova.sav
Data Source Location	Institution: Applied Software Engineering (ISA) research group. University of Seville City: Seville Country: Spain
Data Accessibility	Repository name: EXEMPLAR platform described in [5] Direct URL to data: https://exemplar.us.es/demo/ BernardezMindfulnessTSE
Related Research Article	B. Bernárdez, A. Durán, J. A. Parejo, N. Juristo, A. Ruiz-Cortés, "Effects of mindfulness on conceptual modelling performance: a series of experiments", IEEE Transactions on Software Engineering. DOI 10.1109/TSE.2020.2991699. *In press.*

7.3 Value of the Data

■ The MetaMind dataset provides the key for the correct understanding not only of the meta-analysis, but also of the process followed leading to its generation from the individual experiment datasets.

■ The MetaMind dataset offers the possibility of reproduction of the experimental results. The Association for Computing Machinery (ACM) in [6] clearly states the importance held by experimental results that can be independently reproduced.

■ According to the taxonomy proposed by [6], the MetaMind dataset is labeled as "available" because it has been made permanently available for retrieval. By releasing the dataset in this chapter, we believe that it may be closer to achieving the highest level of replication for the label of "replicated" since, the results of the study have been obtained in a subsequent study by a person or team other than ourselves, having employed, in part, our laboratory package on https://exemplar.us.es/demo/BernardezMindfulnessTSE.

■ Mind#1 already has two internal replications; however, the external replications have to be performed by independent researchers and therefore involve a greater power of confirmation [7]. By releasing the dataset in this chapter, we increase the possibility of external replication, because the laboratory package, including the MetaMind dataset and experimental material, gains visibility. The unavailability of laboratory packages has been identified by the research community as one of the main causes of the low number of replications [8].

7.4 Data Description

The MetaMind dataset is available online as part of a complete laboratory package that also includes files of experimental material and statistical analyses. MetaMind is composed of: (i) the files of the three experiments in the family; and (ii) a set of files for the meta-analysis, which are held in various formats, depending on the tool that is applied to process the data, but all with the same content. Table 7.1 shows the fields of the MetaMind file schema.

Since most of the statistical tests have been performed in both R and SPSS as double-checking, the dataset records in two different formats: csv and sav. As mentioned earlier, although there are two meta-analysis files, they both hold a common content.

On the one hand, *simplifieddata.csv* is the dataset file for the analysis in R-Studio, following the structure in Table 7.1. This file is automatically generated from the individual experiment dataset files (*data2014.csv*, *data2015.csv*, and *data2016.csv*), by selecting, renaming the variables, and adding the information regarding order and year.

Table 7.1   Data Description of the MetaMind Dataset

Subject	An anonymous student identifier.
Efficiency	The score of efficiency for each subject at a particular point in time.
Effectiveness	The score of effectiveness for each subject at a particular point in time.
Year	The year of the individual experiment in which the subject had participated: in the case of Mind#1, the year is *2014*; for Mind#2, the year is *2015*; and for Mind#3, the year is *2016*.
Group	The group to which the subject belongs: either experimental or control. The field has the value *MF* (<u>M</u>ind<u>F</u>ulness) if the subject belongs to the experimental group; otherwise the field has the value *NULL*.
Time	The moment at which the students developed the exercises. If the measure was taken before the treatment (mindfulness workshop) the field has the value *Pre*, otherwise the field has the value *Post*.
Order	The order in which the conceptual modeling exercises were performed in each experiment. Thus, the order in Mind#1 and Mind#2 has the value ERASMUS→EoDP, but in Mind#3 the order has the value EoDP→ERASMUS.
Exercise	The specific conceptual modeling exercise in which the student achieves the corresponding score on the response variables.

On the other hand, by taking into account that, in SPSS, different statistical tests require a different raw-data format, two dataset formats have been considered. First, *fulldata.sav* constitutes the file which has the same structure as *simplifieddata.csv* in that it has 260 rows (two for each subject in the sample). In the second format, the file *fulldataforancova.sav* follows a wide format in that repeated measure data appears in two columns (Pre-Treatment, Post-Treatment) for each response variable.

7.5 Experimental Design, Materials, and Methods

In order to study the evolution of the students over a period of time, a major requirement for choosing the experimental design involved measuring the conceptual modeling performance of each student before and after the period of practice of mindfulness, by means of a repeated measure design. As a consequence, a *within–subjects* factor was identified called *time*, representing the moment in which each conceptual modeling exercises was performed. The factor *time* has two levels, Pre-Treatment and Post-Treatment.

However, the main factor in the family of experiments was a *between–subjects factor*, called *group*, which represents the group in which a particular student was assigned. This factor has two levels: the experimental group (students who practice mindfulness), and the control group (to which a placebo treatment was administered).

Since this design includes both *within* and a *between–subjects* factors, the *2×2 mixed factorial design* proposed by [9] was selected. This design is commonly used in the fields of psychology and medicine when the evolution of patients under a given treatment (or therapy) is studied after a certain amount of time.

The methodology included the following steps: (i) each subject was randomly[2] assigned one single treatment (*group*); (ii) both groups of students developed the first conceptual modeling exercise (Pre-Treatment); (iii) the experimental group practiced mindfulness over a specified period of time; and (iv) both groups of students were subject to the second conceptual modeling exercise (Post-Treatment). Therefore, two repeated measures on the response variables were taken (Pre-Treatment and Post-Treatment).

With respect to the experimental material, two slide presentations were designed: (i) for the recruitment of students; and (ii) for the introduction of students to the practice of mindfulness. Several questionnaires were also used so that the students could express their interest and a demographic study of the sample could be carried out. The statements of the conceptual modeling exercises were written as an interview transcript, that is, they contained questions and answers between a requirement engineer and a customer regarding a particular problem domain and the expected system behavior. As commented above, ERASMUS and EoDP were chosen under the influence of the criteria regarding the familiarity and interest of the students, since these constitute the potential candidates for an ERASMUS grant and they also have to develop an EoDP to finish their degree studies. Furthermore, the experimental material includes the experiment specifications in SEDL (Scientific Experiments Description Language), which is the domain-specific language for the experiment description of the EXEMPLAR platform [5].

In this section the experimental setting of the family of experiment is summarized. A detailed analysis of the threats to the validity of such experimental setting and the actions performed to mitigate them is compiled in [4].

7.6 About Benefits of Mindfulness in Software Engineering Activities

Several Information Technology organizations have successfully incorporated mindfulness into their teams [10, 11]. Among other advantages, the practice of mindfulness can improve the flow of meetings in agile development teams, in terms of effectiveness, decision-making, and listening, as shown in [12].

Although, in our family of experiments, the most influential benefits of mindfulness are attention, concentration, and mental clarity, other proven benefits of mindfulness, such as empathy and emotional intelligence [13], could be explored in areas such as requirement elicitation. In particular, during interviews, requirements engineers should put themselves in the shoes of the client and the user in order to understand their needs with respect to the system to be developed. It is well known that the introverted character of the requirements engineer can sometimes make difficult to achieve the objectives during the interviews [14]. The impact of mindfulness in requirement negotiation or development team management could be also explored.

7.7 Competing Interest

The authors declare that they have no known competing financial interests or personal relationships which have, or could be perceived to have, influenced the work reported in this article.

Acknowledgment

This work has been partially supported by the European Commission (FEDER) and the Spanish Government under projects HORATIO (RTI2018-101204-B-C21), APOLO (US-1264651), and EKIPMENT-PLUS (P18-FR-2895).

Notes

1. Two well-known problem domains were chosen for the conceptual modeling task: the first involved ERASMUS grants, and the second involved End of Degree Projects (EoDP).
2. In Mind#1, the subjects chose whether they wanted to practice mindfulness. In accordance with their selection, they were assigned either to the experimental group or to the control group.

References

1. P. Sedlmeier, C. Loße, and L. C. Quasten, "Psychological effects of meditation for healthy practitioners: an update", Mindfulness, vol. 9, no. 2, pp. 371–387, 2018.
2. B. Bernárdez, A. Durán, J. A. Parejo, and A. Ruiz-Cortés, "A controlled experiment to evaluate the effects of mindfulness in software engineering", International Symposium on Empirical Software Engineering and Measurement, pp. 1–10, 2014.
3. B. Bernárdez, A. Durán, J. A. Parejo, and A. Ruiz-Cortés, "An experimental replication on the effect of the practice of mindfulness in conceptual modelling performance", Journal of Systems and Software, vol. 136, pp. 153–172, 2018.
4. B. Bernárdez, A. Durán, J. A. Parejo, N. Juristo, and A. Ruiz-Cortés, "Effects of mindfulness on conceptual modelling performance: a series of experiments", IEEE Transactions on Software Engineering. *DOI: 10.1109/TSE.2020.2991699*
5. J. A. Parejo, S. Segura, P. Fernández, and A. Ruiz-Cortés, "Exemplar: an experimental information repository for software engineering research", Actas de las XIX Jornadas de Ingeniería del Software y Bases de Datos (JISBD'2014), pp. 155–159, 2014.
6. ACM Publications. "Artifact Review and Badging" [Online]. Available: https://bit.ly/3eFoARy
7. A. Brooks, J. Daly, J. Miller, M. Roper, and M. Wood, "Replication of experimental results in software engineering", Technical Report ISERN-96-10, University of Strathclyde, 1996.
8. M. Cruz, B. Bernárdez, A. Durán, J. A. Galindo, and A. Ruiz-Cortés, "Replication of studies in empirical software engineering: a systematic mapping study from 2013 to 2018", IEEE Access, vol. 8, pp. 26773–26791, 2020.
9. D. T. Campbell and S. Julian, "Experimental and Quasi–Experimental Designs for Research", United States: Wadsworth, 1963.
10. N. Shachtman, "Enlightenment Engineers: Meditation and Mindfulness in Silicon Valley", San Francisco, CA: WIRED Magazine, 2013.
11. Search Inside Yourself Leadership Institute, "Search inside your self–program impact report", 2019. [Online]. Available: https://bit.ly/2LUMNdb
12. P. den Heijer, W. Koole, and C. J. Stettina, "Don't forget to breathe: a controlled trial of mindfulness practices in agile project teams", in Proc. Int. Conf. Agile Soft. Develop., pp. 103–118, 2017.
13. D. Goleman and R. J. Davidson, "Altered Traits: Science Reveals how Meditation Changes Your Mind, Brain, and Body". Penguin, 2017.
14. L. F. Capretz, "Personality types in software engineering", International Journal of Human–Computer Studies, vol. 58, no. 2, pp. 207–214, 2003.

Chapter 8

Process Performance Indicators for IT Service Management: The PPI Dataset

Bedilia Estrada-Torres[a], Adela del-Río-Ortega[a], Manuel Resinas[a], and Antonio Ruiz-Cortés[a]

[a]Universidad de Sevilla, Seville, Spain

Contents

8.1 Background and Summary 125
8.2 Dataset Specification 126
8.3 Value of the Data 128
8.4 Data Description 128
8.5 Experimental Design, Materials, and Methods 128
8.6 Conclusion 130
Acknowledgment 130
References 130

8.1 Background and Summary

In process-oriented organizations, performance measurement is a key factor in determining whether the organization's objectives in a given area are being achieved, or whether corrective actions are needed. Process performance indicators (PPIs) play a relevant role in monitoring the business process and identifying possible improvement areas. They allow the evaluation of efficiency and effectiveness of a given process in a quantitative way using data that is generated within the process flow [1–3].

One of the most important elements of PPIs is their measure definition, which provides all the information required to calculate the PPI. In [4], four performance dimensions are described: time, cost, quality and flexibility; derived from the need for processes to be faster, cheaper, better and amenable to changes. These dimensions are reflected in the PPI measure definition.

Despite the importance of PPIs, it is unusual to find a clearly defined, fully structured set of key organizational indicators. Usually organizations only have documents with poorly structured information or ambiguously defined indicators with incomplete data [2], or also formal definitions that are difficult for non-technical users to understand, such as algorithms or SQL queries [5]. Over time, several proposals have emerged for the definition and modeling of performance indicators [1, 2, 6, 7], including templates [8, 9] and graphical notations [5, 10], as well as, methodologies [11] and analysis techniques, based amongst others on business process intelligence [12], or the use of PPI definitions in natural language processing [13, 14].

However, a common barrier in process performance research is the lack of public datasets of PPIs defined in real contexts. Organizations are reluctant to make them public because they contain sensitive information for the organization or because they provide a competitive advantage. Reference models such as SCOR [15] or ITIL [16] serve as a guide for the definition of metrics but have the disadvantage of providing definitions at a high level, delegating the task of defining and implementing the indicator customizing it to each organization that follows the reference model.

In this chapter, we present the PPI Dataset, a continuously growing public resource containing a set of 102 real performance indicators defined in natural language. These indicators are associated with processes identified in the information technology (IT) service management of several public organizations. The data provided has been previously used in different research approaches, addressing areas as varied as the automatic transformation of natural language PPI definitions into structured PPI definitions that facilitate the automatic calculation of PPIs [13, 14], and the proposal of new indicator management tools and techniques [5].

8.2 Dataset Specification

Subject	Natural Language PPI Definitions
Specific Subject Area	Performance Management
Type of Data	Spreadsheets

How Data were Acquired	The PPI information was obtained from the SLA of the technical specifications of the RFQ that public organizations provided. The information was filtered and structured manually. As the original information was provided in Spanish, the final list of indicators identified was translated into English.
Data Format	PPI definitions are provided in CSV format. Indicator lists in English and Spanish are provided separately. Analyzed. Filtered.
Parameters for Data Collection	We collected the PPIs defined in Service Level Agreements of the technical specifications of the requests for quote that public organizations provided. A total of 102 process performance indicators were identified and defined for several IT service management processes.
Description of Data Collection	The information from the SLAs was filtered and structured to define the indicators. Furthermore, duplicated PPIs were manually removed. The indicators were classified according to the processes to which they are related. After the review of the domain experts, the final list was created. The definitions were translated to facilitate their use in different contexts.
Data Source Location	Institution: Applied Software Engineering (ISA) research group. University of Seville. City: Seville Country: Spain
Data Accessibility	The data is publicly available in a repository. Repository name: The PPI Dataset Direct URL to data: https://github.com/isa-group/ppi-dataset
Related Research Article	del-Río-Ortega, A., Resinas, M., Durán, A., Bernárdez, B., Ruiz-Cortés, A., Toro, M. VISUAL PPINOT: A Graphical Notation for Process Performance Indicators. Business & Information Systems Engineering. 61, 137–161 (2019). https://doi.org/10.1007/s12599-017-0483-3 van der Aa, H., Leopold, H., del-Río-Ortega, A., Resinas, M., Reijers, H. A. Transforming unstructured natural language descriptions into measurable process performance indicators using Hidden Markov Models. Information Systems. 71, 27–39 (2017). https://doi.org/10.1016/j.is.2017.06.005 van der Aa, H., del-Río-Ortega, A., Resinas, M., Leopold, H., Ruiz-Cortés, A., Mendling, J., Reijers, H. A. (2016) Narrowing the Business-IT Gap in Process Performance Measurement. In: Advanced Information Systems Engineering. CAiSE 2016. Lecture Notes in Computer Science, vol 9694. Springer, Cham. https://doi.org/10.1007/978-3-319-39696-5_33

8.3 Value of the Data

- The definitions of indicators contained in the PPI Dataset provide information on the type of performance information that is relevant in commonly used IT service management processes such as those presented in this document (Request for release, Request for change and Incident Management processes).
- Finding explicit definitions of indicators is rare in real contexts, so the definitions presented can be used as a basis for adapting these indicators to other contexts and/or other similar processes.
- PPI definitions are very useful in the field of business process management and can be used in future research to define new PPI definition mechanisms such as templates, automatic indicator calculation tools, modeling techniques and tools, approaches based on process performance analysis using process mining, data structure analysis and/or transformation of information structures based on natural language analysis, among many others.

8.4 Data Description

The PPI Dataset is publicly available on the Github repository: https://github.com/isa-group/ppi-dataset

The repository contains 102 PPIs identified in the service level agreements (SLA) of the technical specifications of the request for quote (RFQ) that several public organizations provided. These indicators are distributed in different files in CSV format, separating the definitions by language (English and Spanish).

- ENG_ThePPIDataset.csv
- SPA_ThePPIDataset.csv

Each CSV file contains the following information:

- *Process:* The process identifier
- *PPI identifier:* A number to identify the PPI
- *PPI name:* Short description of the indicator
- *Calculation of the indicator:* Detailed description indicating how the indicator should be calculated
- *PPI type:* It indicates the type of information or data calculated by the PPI, e.g., time, percentage, etc.

8.5 Experimental Design, Materials, and Methods

The definition of PPIs is associated with processes from which the information to calculate them is extracted. Since these processes were not clearly defined, interviews were conducted with IT staff. With respect to PPIs, the organization provided documentation containing the SLA of the technical specifications of the RFQ.

To obtain the PPI Dataset presented, the following was done:

- From the documents provided, all the information related to the measurement of process performance was extracted.
- As the information was not homogeneous, it was necessary to preprocess the information provided in order to standardize the types of measures, homogenize the vocabulary and eliminate duplicate records.
- PPI definitions were translated from Spanish into English.

In each application scenario where the dataset information has been used, the treatment of the information has been slightly different, depending on the requirements of each case. Three specific scenarios are shown below, which are included in the section "Dataset Specification – Related research article".

In [13], an approach to automatically align natural language descriptions of PPIs to process models is presented. The approach takes as input values a textual description of the PPI and a related process model. It has three steps: (i) the PPIs are classified according to the type of measure; (ii) the textual description of the PPI is analyzed to extract a set of phrases that relate specifically to the parts of the given process, based on a decision tree; and finally, (iii) the results of the previous steps are combined to generate an alignment between the extracted phrases and the elements of the process model.

Similarly, in [14], the data collection was used to automatically translate natural language PPI definitions into definitions according to a structured notation for PPI definition as presented in [8]. The transformation approach (i) analyses a natural language definition of a PPI to identify the parts of that definition that correspond to the parts of the structured definition. To do that, Hidden Markov Models were used. Then (ii) the data required in the structured notation was completed with the information from the previous step to obtain the structured PPI definition.

Finally, as a third scenario, in [9], a graphical notation is proposed for the modeling of PPIs. The dataset presented was used as a basis for graphically modeling PPIs using the VisualPPINOT notation. In addition, it was proved that with the basis for automation provided by the tool that implements the notation, it is possible to calculate the indicators without the need to perform translations to an implementation language, like SQL queries, as is traditionally done to calculate the PPI information.

Although in the examples presented above only the PPI definitions in English were used, in this dataset they are also provided in Spanish, so that they can be used in future research, such as those related to natural language processing.

8.6 Conclusion

In this chapter, we provide a dataset consisting of 102 PPIs defined in natural language. Those PPIs are related to the IT service management processes of several public organizations in Spain.

It is well known that it is difficult to find public datasets that reflect the PPIs used within organizations, as these may reflect sensitive information on the activities carried out within each organization. Therefore, the importance of the dataset provided lies in the fact that it anonymously describes indicators and measures used in real organizations. These PPIs can be used to conduct new research studies or extend existing ones, along lines related to process performance analysis and improvement, natural language processing, or the management of service level agreements, to name a few.

Finally, considering the dynamic nature of organizations, we aim at updating this dataset to reflect possible changes and trends in the way their performance is measured. We also plan to enrich the dataset when more information on other PPIs in new or similar processes becomes available.

Acknowledgment

This work has been partially supported by the European Commission (FEDER), the Spanish Government under projects HORATIO (RTI2018-101204-B-C21) and OPHELIA (RTI2018-101204-B-C22), and the Andalusian R&D&I programs COPAS (P12-TIC-1867), EKIPMENT-PLUS (PR18-FR-2895) and APOLO (US-1264651).

References

1. Popova, V., Sharpanskykh, A. Modeling organizational performance indicators. Information Systems. 4, 505–527 (2010). https://doi.org/10.1016/j.is.2009.12.001
2. del-Río-Ortega, A., Resinas, M., Cabanillas, C., Ruiz-Cortés, A. On the definition and design-time analysis of process performance indicators. Information Systems. 4, 470–490 (2013). https://doi.org/10.1016/j.is.2012.11.004
3. del-Río-Ortega, A., Resinas, M., Ruiz-Cortés, A. Business Process Performance Measurement. Encyclopedia of Big Data Technologies. Springer, Cham. (2018). https://doi.org/10.1007/978-3-319-63962-8_99-1
4. Dumas, M., La Rosa, M., Mendling, J., Reijers, H. A. Fundamentals of Business Process Management. (2018) Springer. https://doi.org/10.1007/978-3-662-56509-4
5. del-Río-Ortega, A., Resinas, M., Durán, A., Bernárdez, B., Ruiz-Cortés, A., Toro, M. VISUAL PPINOT: A graphical notation for process performance indicators. Business & Information Systems Engineering. 61, 137–161 (2019). https://doi.org/10.1007/s12599-017-0483-3
6. González, O., Casallas, R., Deridder, D. MMC-BPM: A Domain-Specific Language for Business Processes Analysis. In: International Conference on Business Information Systems. BIS 2009. Lecture Notes in Business Information Processing, vol 21. Springer, Berlin, Heidelberg. https://doi.org/10.1007/978-3-642-01190-0_14

7. Cuzzocrea, A., Folino, F., Guarascio, M., Pontieri, L. Predictive monitoring of temporally-aggregated performance indicators of business processes against low-level streaming events. Information Systems. 81, 236–266 (2019). https://doi.org/10.1016/j.is.2018.02.001

8. del-Río-Ortega, A., Resinas, M., Durán, A., Ruiz-Cortés, A. Using templates and linguistic patterns to define process performance indicators. Enterprise Information Systems. 10, 159–192 (2016). https://doi.org/10.1080/17517575.2013.867543

9. Saldivar, J., Vairetti, C., Rodríguez, C., Florian, D., Casati, F., Alarcón, R. Analysis and improvement of business process models using spreadsheets. Information Systems. 57, 1–19 (2016). https://doi.org/10.1016/j.is.2015.10.012

10. Friedenstab, J., Janiesch, C., Matzner, M., Muller, O. Extending BPMN for Business Activity Monitoring. Hawaii International Conference on System Sciences, 4158–4167. (2012). IEEE 10.1109/HICSS.2012.276

11. Estrada-Torres, B., Piccoli Richetti, P. H., del-Río-Ortega, A., Araujo Baião, F., Resinas, M., Santoro, F. M., Ruiz-Cortés, A. Measuring performance in knowledge-intensive processes. ACM Trans. Internet Technol. 19 (1), 15 (2019). https://doi.org/10.1145/3289180

12. Delgado, A., Weber, B., Ruiz, F., Garcia-Rodríguez de Guzmán, I., Piattini, M. An integrated approach based on execution measures for the continuous improvement of business processes realized by services. Information and Software Technology. 56 (2), 134–162 (2014). https://doi.org/10.1016/j.infsof.2013.08.003

13. van der Aa, H., del-Río-Ortega, A., Resinas, M., Leopold, H., Ruiz-Cortés, A., Mendling, J., Reijers, H. A. Narrowing the Business-IT Gap in Process Performance Measurement. In: Advanced Information Systems Engineering. CAiSE 2016. Lecture Notes in Computer Science, vol 9694. Springer, Cham. (2016). https://doi.org/10.1007/978-3-319-39696-5_33

14. van der Aa, H., Leopold, H., del-Río-Ortega, A., Resinas, M., Reijers, H. A. Transforming unstructured natural language descriptions into measurable process performance indicators using Hidden Markov Models. Information Systems. 71, 27–39 (2017). https://doi.org/10.1016/j.is.2017.06.005

15. Apics, S.C.C. Supply Chain Operations Reference Model: SCOR Version 11.0. (2015)

16. ITIL Foundation. ITIL 4 Edition. Axelos Books. ISBN: 9780113316076 (2019)

Chapter 9

Prioritization in Automotive Software Testing: Systematic Literature Review and Directions for Future Research

Naohiko Tsuda[a], Ankush Dadwal[a],
Hironori Washizaki[a], Yoshiaki Fukazawa[a],
Masashi Mizoguchi[b], and Kentaro Yoshimura[b]

[a]*Waseda University, Japan*
[b]*Center for Technology Innovation – Controls, Hitachi, Ltd.,*
 Research & Development Group

Contents

9.1   Introduction 134
9.2   Related Work 136
9.3   Methodology...137
      9.3.1   Research Questions...138
      9.3.2   Search and Selection Process...138
              9.3.2.1   Initial Search 139

9.3.2.2 Impurity Removal 140
9.3.2.3 Merge and Duplicate Removal 140
9.3.2.4 Inclusion and Exclusion Criteria..........................140
9.3.2.5 Removal during Data Extraction...........................140
9.3.3 Data Extraction 140
9.3.4 Data Analysis..141
9.4 Results..142
9.4.1 Number of Studies by Year and Publication Type....................142
9.4.2 Applied Research Strategies 142
9.4.3 Testing Methods..143
9.4.4 Quality Assessment..143
9.4.5 Targeted Problems 146
9.4.6 Different Techniques/Tools, Gaps (Originalities), and the
Main Outcomes of the Selected Studies....................................146
9.5 Discussion 147
9.5.1 Answers for RQ1: What Are the Publication Trends for
Prioritization in Automotive Software Testing?..........................147
9.5.2 Answers for RQ2: What Methods Are Used for Prioritization
in Automotive Software Testing? 147
9.5.3 Answers for RQ3: How Are the Studies Distributed Based on
a Quality Evaluation of Prioritization in Automotive Software
Testing? 152
9.5.4 Answers for RQ4: How Does Existing Research on
Prioritization Help Optimize of Automotive Software
Testing? 152
9.6 Threats to Validity..153
9.7 Conclusion 153
Note..154
References 154

9.1 Introduction

Software in the automotive industry is undergoing a major transition. Automakers have been adding new functions and systems to meet market demand for an ever-growing amount of software-intensive functions. However, these new functions and systems have some negative aspects. First, automakers must enhance their testing techniques because vehicle complexity is increasing. Half of all development costs are consumed by testing [1]. While testing a single software system is difficult, testing without prioritization is cumbersome due to the exponential number of products and features. Today, software determines more than 90% of the functionality in automotive systems, and software components are no longer handwritten [2]. Prioritization is a method to rank and schedule test cases.

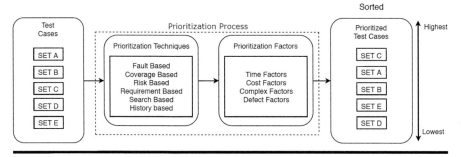

Figure 9.1   General concept of test case prioritization.

It can help enhance the software testing activity. Prioritization in Automotive Software Testing typically decreases the test cases, selects the test cases to be executed, and sets the execution order. Hence, it achieves an early optimization. In this technique, test cases are run in the order of priority to minimize time, cost, and effort during the software testing phase. It also helps identify the best possible combination of a series of test cases that could be executed accordingly. Every organization has its own methods to prioritize test cases. The automotive safety standard ISO 26262 requires extensive testing with numerous test cases. To achieve a high productivity, the availability of quality assurance systems must be high [3].

Figure 9.1 illustrates the general concept of test case prioritization with different sets of test cases (Set A, Set B….Set N). The prioritization process includes both prioritization techniques and factors. The right side shows the prioritized test cases sorted from the highest to lowest.

Herein we use a systematic literature review to evaluate relevant publications on prioritization in the automotive industry. A systematic review aims to assess scientific papers in order to group concepts around a topic. The system review[1] aims to identify common techniques in automotive testing and to define new challenges.

The main contributions of this study are as follows:

■ A reusable framework to classify, compare, and evaluate methods and techniques specific for prioritization in Automotive Software Testing. (Section 9.3)
■ An up-to-date map of prioritization in Automotive Software Testing and its indication for future research. (RQ1, RQ2, and RQ4)
■ An evaluation of the existing research results on prioritization in Automotive Software Testing. (RQ3)

This study should be of interest to both researchers and practitioners aiming to understand existing research on prioritization in Automotive Software Testing and adopt the best solution based on their business goals.

The remainder of this paper is structured as follows:

- Section 2 shows studies related to prioritization in Automotive Software Testing.
- Section 3 details the systematic literature review approach.
- Section 4 presents the results obtained from the systematic literature review.
- Section 5 discusses and statistically analyzes the results.
- Section 6 addresses potential threats to validity.
- Section 7 concludes this paper and presents future work.

9.2 Related Work

The evolving technology for automotive testing impacts automakers. It is nearly impossible to test all systems manually. Automating the testing phase significantly reduces the cost of software development [4], but the literature contains little information about prioritization efforts in the automobile industry. Some publications define general approaches to apply prioritization in software testing, but few studies report the adoption of these techniques into the automotive industry. Herein we focus on known techniques and their applicability to the investigated domain.

In the past few decades, numerous studies have demonstrated that vehicles are becoming increasingly more complex and more connected [2, 3, 5–16]. For example, an empirical study that investigated the potential regarding quality improvements and cost savings, employed data from 13 industry case studies as part of a 3-year large-scale research project. This study identified major goals and strategies associated with (integrated) model-based analysis and testing as well as evaluated the improvements achieved [17].

Rafi et al. [18] (2012) reviewed the benefits and the limitations of automated software testing in the literature. Their review, which included 25 works, tried to close the gap by investigating academic and practitioner views on software testing regarding the benefits and drawbacks of test automation. Although, benefits often come from stronger sources of evidence (experiments and case studies), limitations are more frequently reported in experience reports.

Additionally, their survey of practitioners showed that the main benefits of test automation are reusability, repeatability, and reduced burden in test executions. Of the respondents, 45% agreed that the available tools do not fit their needs, while 80% disagreed with the vision that automated testing would fully replace manual testing.

On the other hand, in the work of [19], the authors presented a systematic literature review of test case prioritization with an emphasis on regression testing between 1999 and 2016. Although many prioritization techniques exist, improvements are still required. They found that prioritization in the early stage of a system development life cycle is worth investigating to ease the burden of project managers and the development team in making the necessary adjustments. Additionally, human involvement in decision-making and estimations needs to be changed into computerized and automated reasoning to reduce human error.

9.3 Methodology

Our systematic literature review began by specifying the scope and searching only documents in the Automotive Software Testing domain. We included studies discussing issues related to prioritization in the field of testing, while excluding topics focusing on software testing without prioritizing the test cases. This research followed the guidelines suggested in [20, 21]. Figure 9.2 shows the four steps of our method: determine the research questions to answer the aim of this study, search relevant repositories, extract data, and analyze the primary studies utilized in the systematic literature review.

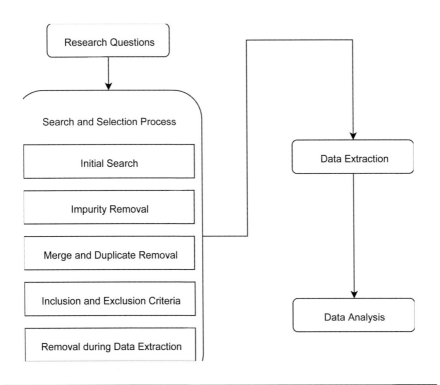

Figure 9.2   Phases of the review protocol.

9.3.1 Research Questions

To understand the abovementioned scenario, we propose the following four research questions.

RQ1: What are the **publication trends** for *prioritization in Automotive Software Testing?*

This research question should characterize the interest and ongoing research on this topic. The goal is to understand the development of test case prioritization in Automotive Software Testing. Additionally, it will identify relevant venues where results are published over time.

RQ2: What **methods are used** for *prioritization in Automotive Software Testing?*

This research question should elucidate the different methods used for prioritization in Automotive Software Testing. The goal here is to determine the main methods and tools employed by researchers. Differences in approaches will provide insight of the performance of each prioritization approach as well as its strengths and weaknesses.

RQ3: How are studies distributed based on a **quality evaluation** of *prioritization in Automotive Software Testing?*

This research question should reveal the quality distribution of the selected primary studies. The goal is to evaluate the selected primary studies.

RQ4: How does **existing research** on prioritization help *optimize Automotive Software Testing?*

This research question should classify existing and future research on prioritization in Automotive Software Testing and assess research characteristics. It will aid researchers and practitioners in selecting which methods are suitable for their controlled experiment or case study. This is the most important and challenging question because it compiles problems that have yet to be solved.

9.3.2 Search and Selection Process

The value of a systematic literature review is generally realized according to the selected primary studies. The search-and-selection process is a multistage process (Figure 9.3). It allows the number and characteristics of the studies that are considered during various stages to be controlled.

As mentioned in [20] and [21], we used three of the largest scientific databases and indexing systems in software engineering:

- IEEE Xplore
- ACM Digital Library
- Scopus

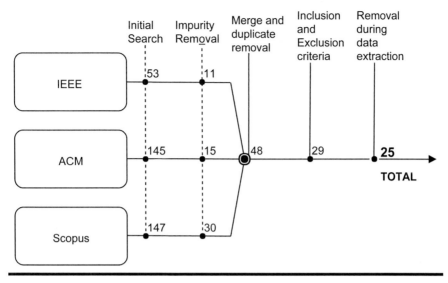

Figure 9.3   Paper selection process.

These online databases were selected because they are common and effective in systematic literature reviews of software engineering. Additionally, the search results can be exported.

Further, these databases provide mechanisms to perform keyword searches. Our search did not specify a fixed time frame and it was very generic (i.e., Prioritization in Automotive Software Testing) to cover as many significant studies as possible.

9.3.2.1 Initial Search

We performed a search in three of the largest and most complete scientific databases and indexing systems in software engineering: ACM Digital Library, IEEE Xplore, and Scopus. These databases were searched using a search string that included important keywords in our four research questions. A systemic method to formulate search keywords must include a process to establish significant terms based on RQs, equivalent words for significant terms, keywords in applicable studies, and usage of the Boolean "OR" and "AND" operators as alternative links between terms. We augmented the keywords with their synonyms, producing the following search string:

```
(("automobile" OR "automotive" OR "car")
AND
("software" OR "program" OR "code")
AND
("prioritization" OR "priority" OR "case selection")
AND
(test*))
```

This combination extracted many relevant studies. For consistency, we executed a query on titles, abstracts, and keywords of papers in all the data sources regardless of time or subject area.

9.3.2.2 Impurity Removal

Due to the nature of the involved data sources, the search results included elements that were clearly not research papers such as abstracts, international standards, textbooks, etc. In this stage, we manually removed these results.

9.3.2.3 Merge and Duplicate Removal

Here all studies were combined into a single dataset. Duplicated entries were matched by title, authors, year, and venue of publication. This process was essential to remove duplicates and unrelated studies.

9.3.2.4 Inclusion and Exclusion Criteria

We considered all selected studies and filtered them according to a set of well-defined selection criteria. The inclusion and exclusion criteria of our study were:

- ■ Inclusion criteria:
  - – Studies focusing on software testing specific to the automotive industry.
  - – Studies providing a solution for prioritizing Automotive Software Testing.
  - – Studies in the field of software engineering.
  - – Studies written in English.
- ■ Exclusion criteria:
  - – Studies focusing on the automotive industry, but do not explicitly deal with software testing.
  - – Studies where software testing is only used as an example.
  - – Studies not available as full text.
  - – Studies not in English.
  - – Studies duplicating of other studies.

9.3.2.5 Removal during Data Extraction

When reviewing primary studies in detail to extract information, all the authors agreed that four studies were semantically beyond the scope of this research. Consequently, they were excluded.

9.3.3 Data Extraction

Relevant information was extracted from the primary studies to answer the research questions. We used data extraction forms to perform this task in an accurate and consistent manner. The data was collected and stored in a spreadsheet using MS

Excel to list the relevant information of each paper. This technique helped extract and view data in a tabular form.

The following information was collected for each paper:

- Publication title
- Publication year
- Publication venue
- Problems faced by the authors
- Testing method used
- Limitations in field
- Detail of the proposed solution
- Results obtained
- Rating of quality issues
- Verification and validation
- Future work suggested by the authors
- Conclusion

Answers to research questions

9.3.4 Data Analysis

Data analysis involved collating and summarizing the data extracted from the primary studies to answer the research questions. The main goal was to understand, analyze, and classify current research on prioritization in Automotive Software Testing. Table 9.1 briefly shows important types of data used to map synthesized data to the research questions.

Table 9.1 Data Collection for Research Questions

Research Questions (RQs)	Type of Data Extracted (and Corresponding Subsections)
RQ1	Number of publications per year, repository, venue (Subsection 9.4.1). Applied research strategies (Subsection 9.4.2).
RQ2	Types of testing techniques and methods, studies, and bibliographic references (Subsection 9.4.3).
RQ3	Distribution of scores per publication and quality evaluation (Subsection 9.4.4).
RQ4	Target problems, techniques/tools, gaps (originalities), and main outcome of the selected studies (Subsections 9.4.5 and 9.4.6)

9.4 Results

9.4.1 Number of Studies by Year and Publication Type

Figure 9.4 presents the distribution of publications over time. Bubble size means the number of studies, where grouped by publication types and years.

The most common publication types are conference papers (17) followed by workshop papers (5), journals (2), and symposiums (1). The first paper related to prioritization in Automotive Software Testing was published in 2008. A small but constant number of publications were published until 2014.

9.4.2 Applied Research Strategies

Table 9.2 shows the applied research strategies. Many papers provide solution proposals (20) and evaluation research (15), indicating that today's researchers focus on industry and practitioner-oriented studies (e.g., industrial case study, action research). Another common research strategy is validation research (14). However, Table 9.2 shows that opinion papers (4) and survey papers (1) are rarely studied.

Table 9.2　Applied Research Strategies

Research Strategies	Number of Studies	Studies
Solution proposal	20	[1–3, 5–8, 11–16, 22–28]
Evaluation research	15	[5, 6, 8–14, 22, 24–26, 28, 29]
Validation research	14	[5–8, 11, 14, 15, 22–26, 28, 29]
Opinion paper	4	[4, 9, 16, 17]
Survey paper	1	[17]

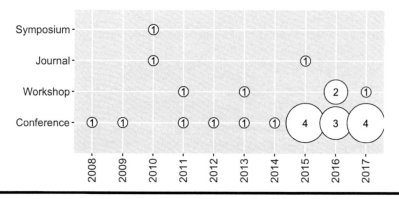

Figure 9.4　Primary studies distributed by type of publication over the years.

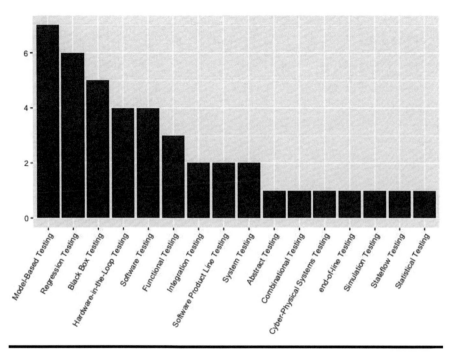

Figure 9.5 Testing methods.

9.4.3 Testing Methods

Figure 9.5 depicts a histogram of the distribution of the most common techniques. These are model-based testing (7), regression testing (6), black-box testing (5), hardware in the loop testing (4), software testing (4), functional testing (3), and other. Model-based testing has gained much interest due to the requirements of connected services for vehicles.

Multiple studies employ different methods or combination of methods (Table 9.3). These papers frequently target model-based testing (7), regression testing (6), and black-box testing (5). Model-based development is an efficient, reliable, and cost-effective paradigm to design and implement complex embedded systems.

9.4.4 Quality Assessment

Study [30] created a ranking to rate papers based on the relevance to the topic and the quality of the paper. Quality Issues (QI) include:

- QI1: Does the paper have clear goals?
- QI2: Does the assessment approach match the goals?
- QI3: Can the method be replicated?
- QI4: Are results shown in detail?

Table 9.3   Applied Testing Methods

Main Testing Methods	Number of Studies	Studies
Model-based testing	7	[1, 2, 7, 8, 17, 22, 28]
Regression testing	6	[2, 3, 5, 7, 13, 22]
Black-box testing	5	[5, 6, 7, 11, 13]
Hardware in the loop testing	4	[6, 11, 16, 29]
Software testing (test case generation based)	4	[4, 16, 25, 27]
Functional testing	3	[11, 26, 29]
Integration testing	2	[8, 15]
Software product line testing	2	[1, 22]
System testing	2	[5, 6]
Abstract testing (authors named)	1	[12]
Combinational testing (authors named)	1	[24]
Cyber-physical systems testing	1	[9]
End-of-line testing	1	[3]
Simulation testing (scenario based)	1	[23]
Stateflow testing	1	[10]
Statistical testing	1	[14]

For each quality issue, the articles were rated as Yes (Y), Partial (P), or No (N).

- Y indicates that the issue is fully addressed in the text.
- P indicates that the issue is partially addressed in the text.
- N denotes that the issue is not addressed in the text.

These ratings were scored as Yes = 1 point, Partial = 0.5, and No = 0. Table 9.4 shows the papers analyzed in this systematic literature review and their respective scores based on the Quality Issues discussed above.

Table 9.4   Rating of Reviewed Articles

Reference	QI1	QI2	QI3	QI4	Total
[1]	Y	Y	Y	Y	4
[2]	Y	P	P	Y	3
[3]	Y	P	Y	P	3
[4]	Y	Y	P	Y	3.5
[5]	Y	P	P	Y	3
[6]	Y	Y	P	P	3
[7]	Y	Y	N	P	2.5
[8]	P	Y	P	Y	3
[9]	Y	P	Y	P	3
[10]	Y	Y	Y	Y	4
[11]	Y	P	Y	Y	3.5
[12]	Y	P	Y	P	3
[13]	Y	Y	P	Y	3.5
[14]	Y	Y	Y	Y	4
[15]	P	P	Y	Y	3
[16]	Y	Y	Y	Y	4
[17]	Y	N	P	P	2
[18]	Y	P	P	Y	3
[19]	Y	Y	P	Y	3.5
[20]	P	N	Y	N	1.5
[21]	Y	Y	Y	Y	4
[22]	Y	Y	Y	P	3.5
[23]	Y	Y	P	Y	3.5
[24]	Y	Y	Y	N	3
[25]	Y	P	Y	P	3
				Average	3.2
	Y(22)P(3)N(0)	Y(14)P(9)N(2)	Y(14)P(10)N(1)	Y(15)P(8)N(2)	

Table 9.5 Target Problems

Main Problems	Number of Studies	Studies
Time consumption	15	[2, 4–16, 23]
Complexity	14	[2, 3, 5–16]
Production cost	13	[2, 5, 6–16]
Test case selection	3	[3, 22, 29]
Quality improvement	3	[17, 26, 28]
Test case generation	2	[25, 27]
Problem space information	1	[1]
Efficiency improvement	1	[24]
Safety conformance	1	[28]

9.4.5 Targeted Problems

Table 9.5 presents the recurring problems targeted by primary studies. Recurring problems include time consumption (15), cost (13), and complexity (14) followed by test case selection (3) and quality improvement (3). To a lesser extent, some studies examine test case generation (2), problem space information (1), efficiency improvement (1), and safety conformance (1).

9.4.6 Different Techniques/Tools, Gaps (Originalities), and the Main Outcomes of the Selected Studies

Table 9.6 presents studies within each category, Techniques/Tools, gaps, and the main study outcomes. Most focus on industrial case studies (n = 14) and technique comparisons (n = 6).

Table 9.6 Validation

Techniques	Number of Studies	Studies
Industrial case study	14	[5, 6, 8, 10–14, 22, 24–26, 28, 29]
Technique comparison	6	[1, 2, 3, 9, 15, 27]
Statistics (only Boxplot)	1	[7]
Test scenario simulation	1	[23]
Others	3	[4, 16, 17]

Table 9.7 lists the techniques used to validate the selected studies. The most common are industrial case studies (14) followed by technique comparisons (6), limited statistical evaluation (1), test scenario simulation (1), and others (3). Technique comparisons include studies that propose and then compare a new method to an old one. Others include three studies, which studies [4, 16] talk about advantages, limitations, and requirements of different approaches. Study [17] is a survey paper from 13 industry case studies.

9.5 Discussion

9.5.1 Answers for RQ1: What Are the Publication Trends for Prioritization in Automotive Software Testing?

Publication trends: The most common study type was solution proposal (n = 20), while the least was survey (1). The amount of publications per year is increasing.

Figure 9.4 demonstrates the scientific interest on prioritization in Automotive Software Testing. Until 2014, a small but constant number of publications were published. However, the number of papers has drastically increased since 2014, demonstrating its increasing importance in research. The high number of conference papers indicates that prioritization of Automotive Software Testing is maturing. Studies published before 2015 refer to slightly different perspectives on prioritization than more recent papers. Studies [1, 2, 8, 17] used model-based testing to improve prioritization by increasing the effectiveness. On the other hand, studies [2, 7] showed potential improvements and proposed new model-based methods.

From the perspective of applied research strategies, current studies can be divided in descending order as follows: solution proposal, evaluation research, validation research, opinion paper, and survey paper (Table 9.2). This highlights the fact that there is some level of evidence (e.g., simulations, experiments, prototypes, etc.) supporting the proposed solutions. However, Table 9.2 also shows that only one study employed surveys (1), suggesting that future studies should fill this area.

9.5.2 Answers for RQ2: What Methods Are Used for Prioritization in Automotive Software Testing?

Methods used: The top three common methods are model-based testing (7), regression testing (6), and black-box testing (5).

The most often common methods used for prioritization can be identified in Automotive Software Testing. Figure 9.5 depicts a histogram of the distribution of the most common techniques in the literature. Studies [1, 2, 7, 8, 17, 22, 28] use model-based testing approaches. Software determines more than 90% of the functionality of automotive systems, and up to 80% of the automotive software can be automatically generated from models [2]. Additionally, model-based testing is a

Table 9.7 Validation, Techniques/Tools, Gaps, and Main Study Outcomes

Validation	Number of Papers	Study	Techniques/Tools	Gaps (Originalities)	Main Outcome(s)
Industrial case study	14 (56%)	[5]	Test case selection based on Stochastic model	Find better heuristic clustering approaches	Regression effort can be minimized
		[6]	Test case selection based on a component and communication model	Change deployed libraries	82.3% reduction in tested functions
		[8]	Taster tool, proposed new framework for MBT	Optimize the classification structure used within the priority assignment procedure	New framework for the MBT
		[22]	Dissimilarity-based TCP	Investigate the fault detection capabilities of the approach	Dissimilarity-based TCP issues are solved
		[10]	Test selection algorithms	Develop optimal guidelines to divide test oracle budget across the output-based selection algorithms	Output-based algorithms consistently outperform coverage-based algorithms in revealing faults
		[11]	Evolutionary testing framework, modularHiL, MESSINA	Investigate how to configure the evolutionary testing system to reduce the number of pre-tests	Solution provide complements systematic testing in that it generates test cases for situations that would otherwise be unforeseen by testers

(Continued)

Table 9.7 (Continued) Validation, Techniques/Tools, Gaps, and Main Study Outcomes

Validation	Number of Papers	Study	Techniques/Tools	Gaps (Originalities)	Main Outcome(s)
		[12]	Migration from traditional to abstract testing	Extend the formalism to handle non-functional requirements	Abstract testing is comparable in effectiveness
		[24]	Equivalence Class Partitioning (ECP), Boundary Value Analysis (BVA), Choice Relationship Framework (CRF)	Investigate efficient test case generation and discover more feasible tools and empirical Studies	Efficient reduction in the final effective number of test cases by 42 (88% reduction)
		[13]	Test case combination	Present six approaches to improve test efficiency	Machine learning approach in black-box testing
		[14]	Combination of test models in MATLAB/ Simulink	NO GAPS MENTIONED	Higher coverage is achieved compared with manually created test cases
		[25]	Automatically generate test cases	Refine the functional coverage model	Improvement actions are identified for test case generation
		[26]	Proposed unified model	Implement a large survey on software specifications in Johnson Controls company	More than 90% of the requirements are represented by the model
		[28]	End-to-end test framework	Investigate the nature of the test model and the relevance to the generated test cases	Automation of executable test script generation

(Continued)

Table 9.7 (Continued)   Validation, Techniques/Tools, Gaps, and Main Study Outcomes

Validation	Number of Papers	Study	Techniques/Tools	Gaps (Originalities)	Main Outcome(s)
		[29]	Search-based testing principles	Further investigate safety requirements	Promising approach to ensure safety requirements
Technique comparison	6 (24%)	[1]	Similarity-Based Product Prioritization w.r.t Deltas	Include more solution space information (such as source code)	Improvement in the effectiveness of SPL testing
		[2]	New model-based method for test case prioritization	Implement performance evaluation with a large-scale case study	Future regression testing can be sped up
		[3]	Combined fault diagnosis and test case selection	Evaluate using an end-of-line test system at a real assembly line	System can find test cases to increase the test coverage
		[9]	Weight-based search algorithms	Study weight tuning, different fitness functions, and cost effectiveness measures	Results suggest that all the search algorithms outperform Random Search
		[27]	Model slicing technique for optimal test case generation	*NO GAPS MENTIONED*	Complexity of Simulink models can be reduced

(Continued)

Table 9.7 (Continued) Validation, Techniques/Tools, Gaps, and Main Study Outcomes

Validation	Number of Papers	Study	Techniques/Tools	Gaps (Originalities)	Main Outcome(s)
		[15]	OUTFIT tool	Evaluate it with other domains such as medical and avionic systems	Potential defects can be effectively identified
Statistical evaluation (Boxplot only)	1 (4%)	[7]	Dependences between the components of embedded systems	Study exhaustiveness of the path search and correctness of path search	Reduction in test-cases for regression testing
Test scenario simulation	1 (4%)	[23]	Parallelly execute loosely coupled segments	Fully automate the process of segmentation and instrumentation	Reduce the simulation testing time for both successful and failed runs
Others	3 (12%)	[4]	Three approaches to automatically generate MC/DC test cases	Focus on MC/DC test case generation from formal specifications	Approaches can be combined to support different kinds of decisions
		[17]	Survey paper	Publish more detailed description of the applied evaluation Approach	Improvements possible with MBAT technologies
		[16]	Study on requirements	NO GAPS MENTIONED	Synect provides easy test requirement specifications and management of the test results

common solution to test embedded systems in automotive engineering. Regression testing is undertaken every time a model is updated to verify quality assurance, which becomes time-consuming as it reruns an entire test suite after every minor change. Test case selection for regression testing after new releases is an important task to maintain the availability [3]. Typical studies focus on black-box testing scenarios because the source code is often unavailable in the automotive domain such as an OEM-supplier scenario [13]. Hardware in the loop testing (4) and test case generation-based software testing (4) are the next most used methods. The most common method of testing the software and the Electronic Control Units (ECU) is the use of Hardware-In-the-Loop (HIL) simulations [16]. Software testing can automatically generate test cases for a software product. Functional testing (3) strives to demonstrate the correct implementation of functional requirements and is one of the most important approaches to gain confidence in the correct functional behavior of a system [29]. Integration testing (2), software product line testing (2), and system testing (2) are used as the time donation when the testing process is limited.

A negative highlight of this systematic review is the fact that only one paper directly employs a scenario-based simulation testing method [23]. If a simulation environment can imitate key criteria of the real-world environment, it should be used to provide early feedback on the vehicles design.

9.5.3 Answers for RQ3: How Are the Studies Distributed Based on a Quality Evaluation of Prioritization in Automotive Software Testing?

Paper quality: All studies present research goals; however, a few studies (2) show unclear results.

Table 9.4 shows the papers analyzed in this systematic literature review and their respective scores based on Quality Issues. Only five of the studies [1, 9, 12, 13, 15] have successfully answered all the quality assessment questions. We found that (15) papers show the results in detail, (8) papers show the partial results, and (2) papers show no results. Most of the papers answer QI1: Does the paper have clear goals?

9.5.4 Answers for RQ4: How Does Existing Research on Prioritization Help Optimize of Automotive Software Testing?

Contents: Almost studies tackled to reduce time consumption, test complexity, and production cost. Most studies (14) were validated by industrial case studies, while only one study (1) by test scenario simulation.

Table 9.5 shows the recurring problems that are targeted by primary studies. Because the testing time is expensive, it should be reduced without an uncontrolled reduction of reliability. The entire test suite must be rerun each time the system is updated or modified. Consequently, the testing process becomes more time-consuming with each modification. Automotive systems are becoming more complex due to the higher rate of integration and shared usage. This high complexity results in numerous interfaces, and many signals must be processed inside the system [7]. Testing activities account for a considerable part of the software production costs. A negative highlight is that only two studies discuss efficiency improvement [24] and safety conformance [28]. Table 9.7 presents the studies within each category, Techniques/Tools, gaps (originalities), and the main study outcomes. Most studies focus on industrial case studies (n = 14) and technique comparisons (n = 6).

9.6 Threats to Validity

As always, threats to validity are unavoidable. The analysis was conducted by a single person. One threat is that some information may be omitted. The main query databases were ACM, IEEE Xplore, and Scopus. This may result in inaccurate query results. Moreover, the analysis is limited by the analytical skills of that single person.

9.7 Conclusion

This study provides an overview of the Prioritization in Automotive Software Testing. Specifically, we performed a systematic literature review on 25 primary studies. The research was conducted by finding, classifying, evaluating, and understanding all of the primary studies. The results should help companies incorporate prioritization into their strategies. Researchers can also benefit because this study depicts the limitations and the characteristics of current research: originality and future work mentioned in each study, and different target problems and validation ways. It will aid researchers and practitioners in selecting what approaches are suitable for their researches.

Obtained findings are as follows:

- (RQ1) Publication trends: The most common study type was solution proposal (n = 20), while the least was survey (1). The amount of publications per year is increasing.
- (RQ2) Methods used: The top three common methods are model-based testing (7), regression testing (6), and black-box testing (5).
- (RQ3) Paper quality: All studies present research goals; however, a few studies (2) show unclear results.
- (RQ4) Contents: Almost studies tackled to reduce time consumption, test complexity, and production cost. Most studies (14) were validated by industrial case studies, while only one study (1) by test scenario simulation.

Future work includes (i) investigating the trends in other embedded and non-embedded domains must be investigated as this should provide a more detailed picture and lessons learned regarding prioritization in Automotive Software Testing, (ii) understanding test execution, test case generation, test case selection, and test analysis through a qualitative study, (iii) conducting a uniform analysis on the difference in the techniques used, and (iv) addressing gaps not filled in identified studies.

Note

1. This paper is an extension of a paper presented at the 6th International Workshop on Quantitative Approaches to Software Quality (QuASoQ) co-located with 25th APSEC 2018 [31].

References

1. M. Al-Hajjaji, S. Lity, R. Lachmann, T. Thm, I. Schaefer, and G. Saake. Delta-oriented product prioritization for similarity-based product-line testing. In 2017 IEEE/ACM 2nd International Workshop on Variability and Complexity in Software Design (VACE), pages 34–40, May 2017. doi: 10.1109/VACE.2017.8.
2. A. Morozov, K. Ding, T. Chen, and K. Janschek. Test suite prioritization for efficient regression testing of model-based automotive software. In 2017 International Conference on Software Analysis, Testing and Evolution (SATE), pages 20–29, Nov 2017. doi: 10.1109/SATE.2017.11.
3. S. Abele and M. Weyrich. A combined fault diagnosis and test case selection assistant for automotive end-of-line test systems. In 2016 IEEE 14th International Conference on Industrial Informatics (INDIN), pages 1072–1077, July 2016. doi: 10.1109/INDIN.2016.7819324.
4. S. Kangoye, A. Todoskoff, M. Barreau, and P. Germanicus. Mc/dc test case generation approaches for decisions. In Proceedings of the ASWEC 2015 24th Australasian Software Engineering Conference, ASWEC '15 Vol. II, pages 74–80, New York, NY, USA, 2015. ACM. ISBN 9781-4503-3796-0. doi: 10.1145/2811681.2811696. URL http://doi.acm.org/10.1145/2811681.2811696.
5. I. Alagz, T. Herpel, and R. German. A selection method for black box regression testing with a statistically defined quality level. In 2017 IEEE International Conference on Software Testing, Verification and Validation (ICST), pages 114–125, March 2017. doi: 10.1109/ICST.2017.18.
6. S. Vst and S. Wagner. Trace-based test selection to support continuous integration in the automotive industry. In 2016 IEEE/ACM International Workshop on Continuous Software Evolution and Delivery (CSED), pages 34–40, May 2016. doi: 10.1109/CSED.2016.015.
7. P. Caliebe, T. Herpel, and R. German. Dependency-based test case selection and prioritization in embedded systems. In 2012 IEEE Fifth International Conference on Software Testing, Verification and Validation, pages 731–735, April 2012. doi: 10.1109/ICST.2012.164.

8. L. Krej and J. Novk. Model-based testing of automotive distributed systems with automated prioritization. In 2017 9th IEEE International Conference on Intelligent Data Acquisition and Advanced Computing Systems: Technology and Applications (IDAACS), volume 2, pages 668–673, Sept 2017. doi: 10.1109/IDAACS.2017.8095175.

9. A. Arrieta, S. Wang, G. Sagardui, and L. Etxeberria. Test case prioritization of configurable cyber-physical systems with weight-based search algorithms. In Proceedings of the Genetic and Evolutionary Computation Conference 2016, GECCO '16, pages 1053–1060, New York, NY, USA, 2016. ACM. ISBN 978-1-4503-4206-3. doi: 10.1145/2908812.2908871. URL http://doi.acm.org/10.1145/2908812.2908871.

10. R. Matinnejad, S. Nejati, L. C. Briand, and T. Bruckmann. Effective test suites for mixed discrete-continuous stateflow controllers. In Proceedings of the 2015 10th Joint Meeting on Foundations of Software Engineering, ESEC/FSE 2015, pages 84–95, New York, NY, USA, 2015. ACM. ISBN 978-1-4503-3675-8. doi: 10.1145/2786805.2786818. URL http://doi.acm.org/10.1145/2786805.2786818.

11. P. M. Kruse, J. Wegener, and S. Wappler. A highly configurable test system for evolutionary black-box testing of embedded systems. In Proceedings of the 11th Annual Conference on Genetic and Evolutionary Computation, GECCO '09, pages 1545–1552, New York, NY, USA, 2009. ACM. ISBN 9781-60558-325-9. doi: 10.1145/1569901.1570108. URL http://doi.acm.org/10.1145/1569901.1570108.

12. F. Merz, C. Sinz, H. Post, T. Gorges, and T. Kropf. Bridging the gap between test cases and requirements by abstract testing. Innovation Systems Software Engineering, 11(4):233–242, December 2015. ISSN 1614-5046. doi: 10.1007/s11334-015-0245-7. URL http://dx.doi.org/10.1007/s11334-015-0245-7.

13. R. Lachmann and I. Schaefer. Towards efficient and effective testing in automotive software development. In Lecture Notes in Informatics (LNI), Proceedings – Series of the Gesellschaft fur Informatik (GI), volume P-232, pages 2181–2192, 2014. URL www.scopus.com. Cited By: 2.

14. S. Siegl, K. S. Hielscher, and R. German. Modeling and statistical testing of real time embedded automotive systems by combination of test models and reference models in matlab/simulink. In 2011 21st International Conference on Systems Engineering, pages 180–185, Aug 2011. doi: 10.1109/ICSEng.2011.40.

15. D. Holling, A. Hofbauer, A. Pretschner, and M. Gemmar. Profiting from unit tests for integration testing. In Proceedings – 2016 IEEE International Conference on Software Testing, Verification and Validation, ICST 2016, pages 353–363, 2016. URL www.scopus.com. Cited By: 1.

16. A. Bansal, M. Muli, and K. Patil. Taming Complexity While Gaining Efficiency: Requirements for the Next Generation of Test Automation Tools. In AUTOTESTCON (Proceedings), pages 123–128, 2013. URL www.scopus.com. Cited By: 1.

17. M. Klas, T. Bauer, A. Dereani, T. Soderqvist, and P. Helle. A large-scale technology evaluation study: Effects of model-based analysis and testing. In Proceedings – International Conference on Software Engineering, volume 2, pages 119–128, 2015. URL www.scopus.com. Cited By: 4.

18. D. M. Rafi, K. R. K. Moses, K. Petersen, and M. V. Mntyl. Benefits and limitations of automated software testing: Systematic literature review and practitioner survey. In 2012 7th International Workshop on Automation of Software Test (AST), pages 36–42, June 2012. doi: 10.1109/IWAST.2012.6228988.

19. M. Khatibsyarbini, M. A. Isa, D. N. A. Jawawi, and R. Tumeng. Test case prioritization approaches in regression testing: A systematic literature review. Information and Software Technology, 93:74–93, 2018. ISSN 0950-5849. doi: https://doi.org/10.1016/j.infsof.2017.08.014. URL http://www.sciencedirect.com/science/article/pii/S0950584916304888.

20. B. Kitchenham and S Charters. Guidelines for Performing Systematic Literature Reviews in Software Engineering, EBSE Technical Report, EBSE-2007-01, 2007.

21. B. Kitchenham and P. Brereton. A systematic review of systematic review process research in software engineering. Information Software Technology, 55(12):2049–2075, December 2013. ISSN 0950-5849. doi: 10.1016/j.infsof.2013.07.010. URL http://dx.doi.org/10.1016/j.infsof.2013.07.010.

22. R. Lachmann, S. Lity, M. Al-Hajjaji, F. Fu¨rchtegott, and I. Schaefer. Fine-grained test case prioritization for integration testing of delta-oriented software product lines. In Proceedings of the 7th International Workshop on Feature-Oriented Software Development, FOSD 2016, pages 1–10, New York, NY, USA, 2016. ACM. ISBN 978-1-4503-4647-4. doi: 10.1145/3001867.3001868. URL http://doi.acm.org/10.1145/3001867.3001868.

23. M. A. A. Mamun and J. Hansson. Reducing simulation testing time by parallel execution of loosely coupled segments of a test scenario. In Proceedings of International Workshop on Engineering Simulations for Cyber-Physical Systems, ES4CPS '14, pages 33:33–33:37, New York, NY, USA, 2007. ACM. ISBN 978-1-4503-2614-8. doi: 10.1145/2559627.2559635. URL http://doi.acm.org/10.1145/2559627.2559635.

24. J. S. Eo, H. R. Choi, R. Gao, S. y. Lee, and W. E. Wong. Case study of requirements-based test case generation on an automotive domain. In 2015 IEEE International Conference on Software Quality, Reliability and Security – Companion, pages 210–215, Aug 2015. doi: 10.1109/QRS-C.2015.34.

25. R. Awedikian and B. Yannou. Design of a validation test process of an automotive software. International Journal on Interactive Design and Manufacturing, 4(4):259–268, 2010. URL www.scopus.com. Cited By: 2.

26. R. Awedikian, B. Yannou, P. Lebreton, L. Bouclier, and M. Mekhilef. A simulated model of software specifications for automating functional tests design. In Proceedings DESIGN 2008, the 10th International Design Conference, pages 561–570, 2008. URL www.scopus.com. Cited By: 1.

27. Z. Jiang, X. Wu, Z. Dong, and M. Mu. Optimal test case generation for simulink models using slicing. In Proceedings – 2017 IEEE International Conference on Software Quality, Reliability and Security Companion, QRS-C 2017, pages 363–369, 2017. URL www.scopus.com.

28. J. Lasalle, F. Peureux, and J. Guillet. Automatic test concretization to supply end-to-end mbt for automotive mechatronic systems. In Proceedings – 2011 International Workshop on End-to-End Test Script Engineering, ETSE 2011, pages 16–23, 2011. URL www.scopus.com. Cited By: 4.

29. F. Lindlar and A. Windisch. A search-based approach to functional hardware-in-the-loop testing. In Proceedings – 2nd International Symposium on Search Based Software Engineering, SSBSE 2010, pages 111–119, 2010. URL www.scopus.com. Cited By: 7.

30. U. Kanewala and J. M. Bieman. Testing scientific software: A systematic literature review. Information Software Technology, 56(10):1219–1232, October 2014. ISSN 0950-5849. doi: 10.1016/j.infsof.2014.05.006. URL http://dx.doi.org/10.1016/j.infsof.2014.05.006.

31. A. Dadwal, H. Washizaki, Y. Fukazawa, T. Iida, M. Mizoguchi, and K. Yoshimura. Prioritization in automotive software testing: Systematic literature review. In Horst Lichter, Thanwadee Sunetnanta, Toni Anwar, and Taratip Suwannasart, editors, Proceedings of the 6th International Workshop on Quantitative Approaches to Software Quality co-located with 25th Asia-Pacific Software Engineering Conference (APSEC 2018), Nara, Japan, December 4, 2018, volume 2273 of CEUR Workshop Proceedings, pages 52–58. CEUR-WS.org, 2018. URL http://ceur-ws.org/Vol-2273/QuASoQ-07.pdf.

Chapter 10

Deep Embedding of Open Source Software Bug Repositories for Severity Prediction

Abeer Hamdy[a] and Gloria Ezzat[a]

[a]*British University in Egypt*

Contents

10.1	Introduction	160
	10.1.1 Aims and Contributions	161
	10.1.2 Research Questions	161
10.2	Background	162
	10.2.1 Bug Report and Bug Life Cycle	162
	10.2.2 Word Embedding	163
	10.2.3 Deep Learning	164
	10.2.4 Convolutional Neural Networks (CNNs)	164
	10.2.5 Recurrent Neural Networks (RNNs)	165
	10.2.6 Long Short-Term Memory (LSTM)	166
	10.2.7 Gated Recurrent Unit (GRU)	167
10.3	Methods and Tools	167
	10.3.1 Preprocessing	167
	10.3.2 Embedding	168
	10.3.3 Deep Learning Architectures	169
	10.3.3.1 CNN Architecture	169

        10.3.3.2   LSTM/GRU Architectures.....................................169
        10.3.3.3   Hybrid CNN-LSTM/GRU Architectures................170
10.4   Experiments 170
        10.4.1   Datasets...170
        10.4.2   Evaluation Metrics..170
        10.4.3   Results...171
10.5   Related Work 174
10.6   Conclusion 175
References 176

10.1 Introduction

Bug tracking systems (e.g., Bugzilla and Jira) are utilized by large scale software projects in the maintenance phase to store bug information and fixes 1, 2.. Such systems include repositories to hold bug reports submitted by developers and users of the software. The bug report includes a textual description to the bug in addition to some other attributes. When a bug report is submitted, it is analyzed by a person (triager) who decides whether there is a duplication of that bug or not, then determines its severity, priority and assign it to a developer to be fixed. The severity of a bug is an indicator to its impact on the system and is one of the factors that influence the priority of the bug. High priority bug should be resolved immediately while low priority bug could be fixed in the forthcoming releases [1.. Therefore, severity is considered one of the bug report's crucial attributes. Bugzilla has seven levels to assess the severity of a bug; three levels indicate a bug with high severity ("Blocker", "Critical", "Major"); two levels indicate low severity bug ("Minor", "Trivial"); in addition to the "Enhancement" and "Normal" levels.

The enormous number of daily submitted bug reports makes it hard to ensure an accurate manual triaging process [2]. For instance, Mozilla has received 170 new bug reports in the months from January to July 2009, while Eclipse has received 120 [3]. In Addition, Xia et al. [4] has observed that almost 80% of the reported bugs have their attributes (including the "severity") amended. Likewise, they demonstrated that the reports with amended fields need an extra time to be resolved compared to unamend ones. Hence, there is a compelling need to automate the process of severity identification, in order to speed it up and enhance its accuracy. Consequently, improve the quality of software maintenance phase. Several approaches have been introduced in the literature to automate the assignment of severity levels [1, 2, 5–17]. Early approaches are based on modeling the bug reports using the bag of words (BOW) models [18, 19]; then classifying them into different severity levels using one of the traditional classification techniques (e.g., Naïve Bayes, Support vector machine, etc.). However, the BOW models deal with the bug report as a bag-of-words and do not consider the semantic similarity among synonymous words. Moreover, BOW model suffer from the high dimensionality and sparse data [20, 21].

Recently, neural network models were proposed to learn semantic similarity between words [22]. These models rely on distributional hypothesis, which states that words in the same context tend to have similar semantic meanings. In these models each word is represented by a vector of real numbers that capture their contextual semantic meanings, such that similar words have similar vector representations; this vector is called word embedding. Furthermore, recurrent neural network (RNN) deep learning approaches [23–25] have emerged as a powerful mechanism for providing a sentence representation to a sequence of words modeled using their word embedding vectors. This lead to a breakthrough in the domain of natural language processing, e.g., text classification, language translation, clone detection and code smells identification [26–30].

10.1.1 Aims and Contributions

This chapter leverages the capabilities of both the deep learning and word embedding approaches for mining software bug repositories; in order to identify the severity level of a newly submitted bug report. In this chapter we utilized five deep learning architectures which are: (1) two recent variants of RNNs which are: Long short term memory (LSTM) and Gated recurrent units (GRU), (2) Convolutional neural network (CNN) [31–33] and (3) hybrid CNN-LSTM and CNN-GRU architectures.

The motivation for the proposed approach is the following:

1. The low accuracy of predicting some severity classes in the literature could be attributed to one or more of the following reasons: (a) the information retrieval model used to represent the bug reports, (b) the classifier utilized, (c) the uneven distribution of the bug reports over the severity classes (repository imbalance). In this chapter we focus on the influence of the first two factors.
2. Embedding models provide a robust method for representing text documents in general, due to their ability to acquire semantic relations among words and sentences.
3. CNNs are capable of extracting the key features of an input text and reducing the size of the feature space; which leads to a higher classification accuracy [32, 33].
4. LSTMs and GRUs are capable of providing representations to sequence of words (sentences); however, GRU cell has simpler structure than LSTM cell.
5. Hybrid CNN-RNN architectures could benefit from the capabilities of both the CNNS and RNNS.

10.1.2 Research Questions

We aim at answering the following research questions:

RQ1: How effective are the proposed deep learning architectures in mining the bug repositories for the purpose of identifying the severity levels of newly submitted bug reports?

RQ2: How much improvement could the deep learning architectures achieve over the traditional classifiers trained on the embedding vectors of the bug reports?

RQ3: How much improvement could the superior deep learning model achieve over some state-of-the-art studies?

The chapter is organized as follows: Section 10.2 introduces the background; Section 10.3 presents the proposed methodology; Section 10.4 discusses the experiments and the results; Section 10.5 briefs the related work. Finally, Section 10.6 concludes the chapter and discusses further extensions.

10.2 Background

10.2.1 Bug Report and Bug Life Cycle

A bug report is a software artifact that describes a reported bug of a software project. Figure 10.1 is a sample Eclipse bug report number 446215, which shows the essential components of a bug report. The report includes ID, summary, status, description, comments, attachment, importance, reporter, and fixer. As could be noted, this bug report was submitted by "Marcel" and was assigned to a developer named "Daniel" to be fixed. The bug is related to an Eclipse product "EPP", and the status of this bug is "CLOSED" and "FIXED". The bottom of the report includes the bug description. Two fields in the bug report that denote the importance of the bug: the "Priority" and "Severity" values for bug number: 446215 are "P1" and "Normal". Fixing priority P1 denotes the highest priority and means that the bug should be scheduled for fixing as early as possible. Severity represents the

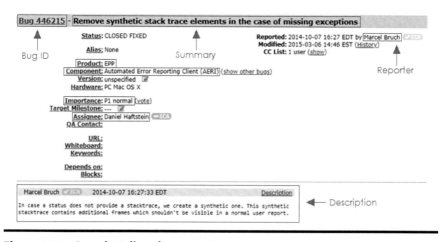

Figure 10.1   Sample Eclipse bug report.

degree of the impact of the bug on the operation of the system. It includes seven degrees (Blocker, Critical, Major, Normal, Minor, Trivial, Enhancement); Blocker bugs are the highest severity bugs; they may block the development or the testing progress. While reports marked as Enhancements are not bugs, they are requests for new functions. Critical bugs may cause program crashes, loss of data or severe memory leaks. A Major bug results in a major loss of functionality. A Normal bug is a regular bug which may cause loss of functionality under specific circumstances; this is the default severity level option in Bugzilla. A Minor bug causes minor loss of functionality. Finally, Trivial bugs are generally superficial issues such as mis-spelled words.

When a bug report is submitted, its status field is labeled "Unconfirmed". Next, a person titled "triager" confirms its presence and changes the status to "New". Then, determines the severity level of the bug, assigns it to a developer for fixing, and changes the report status to "Assigned". If the developer fixes the bug, the status of the bug will be changed to "Fixed" or "Resolved". Finally, the triager confirms whether the bug is fixed, and labels its status as "Closed"; if the bug fix-ing task is finished. However, if the bug is not fixed completely, it may be reopened again, by the assignee to try to fix it or by the triager and reassigned to another developer. In this situation, the bug status changes to "Reopened". In the experi-ments conducted in this work, we considered only the bug reports labeled "Fixed" and "Closed" because they are reliable reports.

10.2.2 Word Embedding

Embedding (also known as distributed representation [22, 34]) is a technique for learning the representation of entities (words or sentences) as real-valued, fixed-length vectors in a high-dimensional space. The vectors are learned such that the entities that have similar meanings are close to each other in the vector space [23]. Embedding provide a more expressive representation for text than classical methods like bag-of-words model, where semantic similarity between words or tokens are ignored, or considered through using n-grams. Word embedding (or Word2vec) is the vector representation of words. Mikolov et al. [22] proposed two techniques for Word2vec namely "Continuous Bag of Words" (CBOW) and "Skip-Gram" (SG). Where, a neural network that captures the relations between a word and its contextual words is built [22]. CBOW learn word rep-resentations that maximize the classification of the current word based on the context words in the same sentence. While SG learns word representations that maximize the classification of the surrounding words based on the current word in the same sentence.

Word2vec is extended for embedding sentences/paragraphs as vectors [35–37]; such that the embedding vectors for sentences of similar meaning are as close as possible, while the vectors for sentences of different meanings are as far apart as possible. One of the sentences embedding techniques (Sent2vec or Doc2vec) is the

Paragraph Vector - Distributed Bag of Words (PV-DBOW) which is an extension to the Word2vec SG model [37]. In this model the neural network is trained to predict the probability distribution of words in the given paragraph based on the words in the paragraph. The second algorithm is Paragraph Vector distributed memory (PV-DM) which is an extension to CBOW model [35].

10.2.3 Deep Learning

Deep learning is a recent field in machine learning that relies on layers of artificial neural networks to learn representations of data with multiple levels of abstraction [38]. Deep learning architectures have been used extensively for addressing a multitude of classification, and prediction problems. The following subsections introduce the widely used deep learning architectures in the literature.

10.2.4 Convolutional Neural Networks (CNNs)

CNNs [31] are inspired by the hierarchical organization of the visual cortex and have been proven effective for text classification [32, 33] same as they are effective for image analysis. In text classification applications the text is supplied to the CNN using one of two forms. The first form is a 1D vector obtained from a model like the BOW model. While the other form is the 2D matrices obtained from one of the word embedding models.

To explain how CNN is used for text classification, let's consider a sentence of length l words that is represented using a 1D vector as follows:

$$x_{1:l} = x_1, x_2 \ldots \ldots x_l \tag{10.1}$$

where, $x_{i:j}$ is a window of words starts from word i to word j; x_i is a value represents the weight/importance of the word i to the sentence $x_{1:l}$. The convolution operation involves applying a filter to windows of size n words to produce new features. Assume a new feature c_i that is generated from applying a filter over a window of words $x_{i:i+n-1}$ will be calculated using Equation 10.2:

$$c_i = f\left(w.x_{i:i+n-1} + b\right) \tag{10.2}$$

where, f is a non-linear function and b is a bias term. The filter is applied to every possible n words window. Consequently, a feature map C of size $(l - n + 1)$ is produced, and defined as follows:

$$C = \left\{c_1, c_2, \ldots \ldots c_{l-n+1}\right\} \tag{10.3}$$

A max-pooling operation is then applied to the feature map C to capture the key features. So, the max-pooling layer reduces the dimensionality of the feature map;

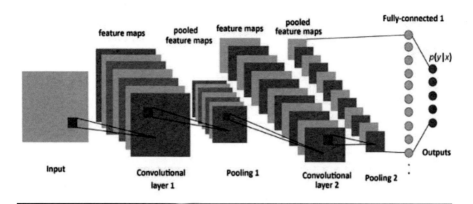

Figure 10.2   A sample CNN network.

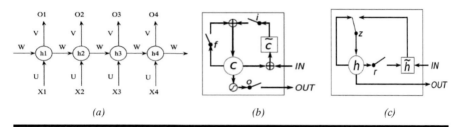

(a)   *(b)*   *(c)*

Figure 10.3   (a) An unrolled RNN neural network unit, (b) LSTM unit, and (c) GRU unit.

consequently, it is considered a feature selection operation. Usually, the CNN utilizes multiple filters with varying window sizes (kernels) to produce different features.

Finally, the output of the max-pooling layer is passed to a dense Softmax layer, whose output is probabilities of the class labels, Figure 10.2 depicts a sample CNN network encompasses two convolutional layers.

10.2.5 *Recurrent Neural Networks (RNNs)*

RNNs [23–25] are other deep learning architectures which have been proven effective in processing sequential data like text; due to their capability to dynamically "memorize" information provided in previous states and incorporate it to a current state. The RNN computational units are connected in a directed cycle such that at each time step *t*, each unit gets two inputs: (i) the current time step X_t, (ii) the hidden state of the same unit from previous time step h_{t-1}, and returns the new hidden state h_t, as depicted by Figure 10.3(a); h_t is calculated using Equation (10.4).

$$h_t = f\left(X_t, h_{t-1}\right) \qquad (10.4)$$

where, h_0 is usually initialized as a vector of *Zeros* and X_i could be a 1D or 2D vector. f is a recursive function; the simplest recursive function is implemented using Equation 10.5 as follows:

$$h_t = tanh\left(W_x X_t, U_h h_{t-1}\right) \tag{10.5}$$

where, W_x, U_h are weight matrices.

However, the simple recursive function given by Equation 10.5 suffers from the problem of vanishing gradient, so more complicated recursive functions were recommended in the literature, e.g., LSTM [24] and GRU [25].

10.2.6 Long Short-Term Memory (LSTM)

LSTM is one of the recent variants of RNN, which is able to preserve long-term dependencies. Furthermore, LSTM has been found to perform reasonably well on various data sets within the context of applications that exhibit sequential patterns, such as: text classification, language translation and source code clone detection [26–29]. The LSTM computational unit comprises a memory cell and three gates: (i) an input gate, (ii) an output gate and (iii) a forget gate, as depicted by Figure 10.3(b). The three gates control the flow of information into and out of the cell; where, each gate is composed of a sigmoid layer and a pointwise multiplication operation. This structure enables the LSTM to overcome the vanishing gradient problem. The inputs to the LSTM unit at time step t are: h_{t-1}, c_{t-1}, c_t, the outputs are: c_t, h_t and they are updated by Equations 10.6–10.11 as follows:

$$i_t = \sigma\left(W_i X_t + U_i h_{t-1} + b_i\right) \tag{10.6}$$

$$o_t = \sigma\left(W_o X_t + U_o h_{t-1} + b_o\right) \tag{10.7}$$

$$f_t = \sigma\left(W_f X_t + U_f h_{t-1} + b_f\right) \tag{10.8}$$

$$\tilde{c}_t = tanh\left(W_c X_t + U_c h_{t-1} + b_c\right) \tag{10.9}$$

$$c_t = \left(i_t \odot \tilde{c}_t\right) + \left(f_t \odot c_{t-1}\right) \tag{10.10}$$

$$h_t = o_t \odot tanh\left(c_t\right) \tag{10.11}$$

where, i_t, f_t, o_t are the input, forget and output gates. $\tilde{c}_t$, c_t are the candidate and new memory cell content. h_t is the activation. σ is the logistic sigmoid function, $\odot$ is the pointwise vector multiplication. h_0, c_0 are usually initialized to zeros.

10.2.7 Gated Recurrent Unit (GRU)

GRU is a variant to the LSTM which has simpler structure, yet is still able to preserve long-term dependencies. Furthermore, GRU has achieved competitive performance with LSTM for many natural language tasks [25, 26].

GRU computational unit merges the LSTM forget and input gates into a single gate, and merges the memory cell state and hidden state, in addition to other changes, as depicted by Figure 10.3(c). The GRU unit output at a time step t is updated using Equations 10.12–10.15 as follows:

$$r_t = \sigma\left(W_r X_t + U_r h_{t-1}\right) \tag{10.12}$$

$$z_t = \sigma\left(W_z X_t + U_z h_{t-1}\right) \tag{10.13}$$

$$\widetilde{h}_t = \tanh(W_h X_t + U_h\left(r_t \odot h_{t-1}\right)) \tag{10.14}$$

$$h_t = \left(1 - z_t\right) \odot h_{t-1} + z_t \odot \widetilde{h}_t) \tag{10.15}$$

where, r_t, z_t are the update and the reset gates; h_t, $\widetilde{h}_t$: are the activation and the candidate activation.

So, there is no decisive conclusion which deep learning model is the best for the text classification task. This is the reason; in this chapter we tackled the severity identification problem with five different deep learning architectures.

10.3 Methods and Tools

The task of fine-grained severity identification could be formulated as a supervised multi-classification problem with input data consists of the summary and description sections of bug reports. While, the output is one of the five severity levels (classes). The proposed methodology (as depicted by Figure 10.4) starts with preprocessing the bug reports, then learning the word embedding vectors which are supplied to the deep learning architecture. The following subsections discuss these phases in details.

10.3.1 Preprocessing

Both of the summary and description sections of each bug report were concatenated, while the rest of the report's fields were discarded. We followed the preprocessing steps discussed in [15] to yield the key words of the corpus of reports. The rarely occurring words were removed to reduce the vocabulary size; we considered only the words occurring with a minimum frequency of 5. NLTK library [39] was used for the preprocessing.

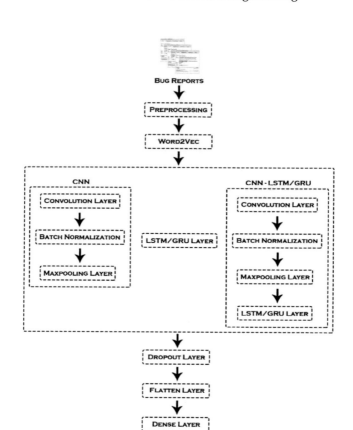

Figure 10.4   Proposed deep learning architectures.

10.3.2 Embedding

Given the preprocessed textual content of the bug reports, the words are mapped into fixed length vector representations using the Word2vec embedding approach, and then supplied to the deep learning architectures. The embedding vectors were trained using their domain corpus (the corresponding software repository). We experimented with the CBOW and SG models, and vector dimensionality of values equal to: 200, 500 and 700. Words that do not present in the set of pre-trained words are initialized randomly. Furthermore, we trained Doc2vec models where each bug report is represented using a fixed length vector. The bug reports' embedding vectors are supplied to three traditional machine learning techniques, which are: KNN, NB and SVM, to compare the performance of the traditional machine learning techniques with the proposed deep learning architectures. Gensim library [40] was used for training both of the Word2vec and Doc2vec models.

10.3.3 Deep Learning Architectures

The following subsections explain the configuration of each of the proposed five deep learning architectures.

10.3.3.1 CNN Architecture

The CNN architecture utilized in this chapter includes two main modules: a feature extraction module followed by a classification module. The feature extraction module is composed of a convolution layer followed by a batch normalization and a max pooling layers. The convolution layer performs convolution operations based on the specified filter and kernel parameters and computes accordingly the network weights to the next layer. The Kernel size was set to equal 3, 5 and 7; while the number of filters was set to equal 8, 16 and 32. We set different values to the hyperparameters in order to experiment with different configurations of the architecture.

The max pooling layer reduces the dimensionality of the feature space by half. The pooling window size was set to 3 and 5. The batch normalization layer optimizes the training through mitigating the impact of the varied input distributions for each training mini-batch [41]. We experimented with one, two and three CNN layers. Multiple CNN layers were used so the model can learn the long-term dependencies.

The output of the max pooling layer is fed to a dropout layer. The dropout layer performs another type of regularization by ignoring some randomly selected nodes during training to prevent over-fitting [42]. The dropout rate was set to equal 0.2... The output of the last dropout layer is fed to a flatten layer then the classifier module. The classifier module consists of two dense layers. The activation function of the first dense layer was set to be Relu While, the activation function of the second dense layer was set to Softmax, and the loss function was set to the Categorical cross-entropy function.

10.3.3.2 LSTM/GRU Architectures

Each of these two architectures consists of a stack of LSTM or GRU layer, dropout layer then a dense layer (same as the dense layer described in the previous subsection). The LSTM/GRU layer learns the sentence (bug report) representation. One of the advantages of the LSTM/GRU is that they can effectively handle sequences of varying lengths. We set the regular dropout (dropout layer) to 0.5, while the recurrent dropout parameters of LSTM layer to 0.1. The recurrent dropouts drop the connections between the recurrent units along with dropping units at inputs and/or outputs [26].

As in the case of the CNN architecture, we experimented with one and two LSTM/GRU layers. We tried different values for the model's hyper-parameters, and we recorded in this chapter the best results achieved.

10.3.3.3 Hybrid CNN-LSTM/GRU Architectures

In this architecture we stack the CNN and the LSTM/GRU layers into a single architecture, so the model can learn the long-term dependencies in the bug reports more efficiently. The CNN layer learns to extract the high-level features (due to the locality of the convolution and pooling), and then the LSTM/GRU layer learns the long-term dependencies. The Keras [43] library was used for implementing all the deep learning models.

10.4 Experiments

This section introduces the datasets, metrics used to evaluate the effectiveness of the proposed models, and the results.

10.4.1 Datasets

Our experiments were conducted on two large repositories: Mozilla and Eclipse; Table 10.1 lists the characteristics of the two datasets. The datasets include the Fixed and Non-duplicated bug reports of 12 products of Mozilla (Firefox, content services, cloud services, etc.) and 19 products of Eclipse (APT, Core, Platform, JDT, etc.). We solely considered in our experiments the bug reports which are labeled "Fixed" and "Closed" as they are reliable. Moreover, we discarded the bug reports whose severity levels assigned as "Enhancement" or "Normal", Likewise, in [6, 7, 13–15]; as Enhancement bug reports are not real bugs but are requests for new features. While the Normal severity level is the default level in Bugzilla and a bunch of reports may have been wrongly assigned.

10.4.2 Evaluation Metrics

As bug severity identification is a multi-class classification problem, we used three widely used metrics which are: Precision, Recall and F-measure. Each metric is

Table 10.1 Characteristics of the Eclipse and Mozilla Bug Reports

Software Repository	Total Number of Reports	Number of Enhancements Reports	Number of Normal Severity Reports	Number of Valid Reports	Number of Bug Reports Per Severity level				
					Blocker	Critical	Major	Minor	Trivial
Mozilla	17,916	850	14,443	2,623	283	508	696	692	444
Eclipse	41,151	4,241	29,555	7,355	728	1,435	2,977	1,383	832

calculated for each severity class. Assume the Blocker class, Precision, Recall, and F-measure are calculated using Equations 10.16–10.18.

$$\text{Precision} = \frac{\text{Number of Blocker reports correctly classified}}{\text{total number of reports classified as Blocker}} \tag{10.16}$$

$$\text{Recall} = \frac{\text{Number of Blocker reports correctly classified}}{\text{total number of Blocker reports}} \tag{10.17}$$

$$\text{F}-\text{measure} = 2 \times \frac{\text{Precision} \times \text{recall}}{\text{Precision} + \text{Recall}} \tag{10.18}$$

10.4.3 Results

Answer to RQ1: Table 10.2 and Figure 10.5 depict the performance of the five deep learning architectures over Eclipse and Mozilla repositories. As could be observed

Table 10.2   Comparison among the Performance of the Proposed Deep Learning Architectures

	Severity	CNN-LSTM			LSTM			CNN-GRU			GRU			CNN		
		P%	R%	F%	P%	R%	F%	P%	R%	F%	P%	R%	F%	P%	R%	F%
Eclipse	Blocker	35	21	26	50	06	11	29	18	23	34	14	20	40	22	29
	Critical	44	31	36	49	27	35	51	20	29	46	28	35	50	22	31
	Major	51	69	59	47	85	61	49	75	59	50	71	59	50	75	60
	Minor	43	44	43	47	28	35	46	39	42	41	42	42	46	31	43
	Trivial	36	20	26	33	10	16	32	23	27	31	23	26	38	27	32
	Weighted Micro-Average	45	47	44	46	47	41	45	46	43	44	46	43	47	48	45
Mozilla	Blocker	79	49	61	80	41	54	53	51	52	60	36	45	80	51	62
	Critical	42	55	48	45	34	39	51	43	46	43	55	49	55	47	51
	Major	35	38	36	35	53	42	38	45	41	37	47	41	41	58	48
	Minor	40	41	41	36	48	41	40	43	41	43	38	41	57	41	48
	Trivial	29	20	24	26	07	11	30	23	26	29	19	23	35	42	38
	Weighted Micro-Average	41	40	40	41	39	37	41	41	41	41	40	40	51	48	48

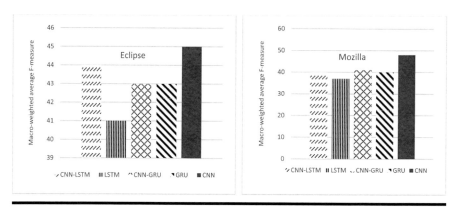

Figure 10.5 Comparison among the micro-average F-measure of the proposed deep learning architectures.

the CNN model achieved the highest weighted macro-average F-measure over each of the Eclipse (45%) and Mozilla (48%). While LSTM has the lowest weighted macro-average F-measure (41%) although LSTM and GRU are considered the ideal solutions for text classification problems. It is noteworthy that GRU is superior to LSTM by 2% in Eclipse and 3% in Mozilla although its architecture is simpler than the LSTM. However, integrating the CNN with the LSTM boosts the average performance of the LSTM by 3% in Eclipse and 1% in Mozilla. On the level of each severity class the CNN achieved the best performance in Mozilla. While, in Eclipse, GRU, LSTM and CNN-LSTM surpassed CNN over the Critical class. It could be observed also that all the deep learning models are biased toward the Major class in Eclipse (highest classification performance) as Eclipse repository is highly imbalanced; the Major class represents about 40% of the repository.

Answer to RQ2: We selected three traditional machine learning techniques (KNN, NB and SVM) that proved their effectiveness in the literature; Scikit learn [44] was used for the implementation. However, in our experiments we did not use the traditional KNN model, we utilized the weighted KNN model utilized in [14], *k* was set to equal 5. Table 10.3 lists the classification performance of the three traditional classifiers against the classification performance of the CNN model. The input to each of the traditional models is the Doc2vec embedding vectors. We have tried different parameters to the SVM; the best results are listed in Table 10.3. As could be observed the SVM could achieve the weighted macro-average F-measure performance of the CNN in Eclipse. While the KNN surpassed the CNN in Mozilla (Figure 10.6).

Answer to RQ3: We compared the results of the CNN model to four previous models [7, 8, 15]. We implemented the models using the settings discussed by the authors. Table 10.4 lists the results over the Eclipse and Mozilla datasets. As observed the CNN model surpasses all the previous models on the level of each severity class except the trivial class. Both of the NB [8] and the NBM [7] surpassed

Table 10.3   Comparison among the Performance of the CNN Architecture and Three Traditional Machine Learning Models Trained on the Doc2vec Features

	Severity	CNN			Doc2vec+KNN K=5			Doc2vec+SVM			Doc2vec+NB		
		P%	*R%*	*F%*	*P%*	*R%*	*F%*	*P%*	*R%*	*F%*	*P%*	*R%*	*F%*
Eclipse	Blocker	40	22	29	34.8	24.9	29	55	14	22	23	53	32
	Critical	50	22	31	30.3	22	25.4	60	23	33	41	32	36
	Major	50	75	60	48.7	59.2	53.4	48	88	62	58	30	39
	Minor	46	31	43	38.9	41.3	40.1	57	41	48	35	50	41
	Trivial	38	27	32	35.6	30.4	32.8	50	13	21	26	33	29
	Weighted Micro-Average	47	48	45	40.4	41.9	40.7	53	51	45	43	37	37
Mozilla	Blocker	80	51	62	82.6	48.8	61.3	79	26	39	58	48	53
	Critical	55	47	51	77.4	45.9	57.6	62	47	54	33	45	38
	Major	41	58	48	42.6	51.9	46.8	42	65	51	42	23	30
	Minor	57	41	48	44.1	60	50.8	46	61	52	38	35	36
	Trivial	35	42	38	39.9	32.9	36	33	12	17	28	43	34
	Weighted Micro-Average	51	48	48	53.6	49.3	49.7	49	47	45	39	36	36

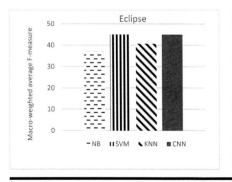
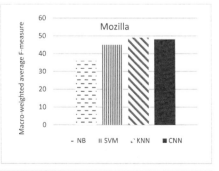

Figure 10.6   Comparison among the micro-average F-measure of the CNN architecture and the traditional machine learning models trained on the Doc2vec embedding vectors.

Table 10.4 Comparison among the Performance of the CNN Architecture and Three State-of-the-Art Models

| | | CNN | | | Hamdy and El-Laithy [15] | | | Chaturvedi and Singh [8] | | | | | | Lamkanfi [7] | | |
| | | | | | | | | KNN: K=3, TF-IDF, 125 Unigrams | | | TF-IDF-NB | | | TF-IDF-NBM | | |
		P%	R%	F%	P%	R%	F%	P%	R%	F%	P%	R%	F%	P%	R%	F%
Eclipse	Bl	40	22	29	33.5	23.2	27.4	17.9	27.2	21.6	23.3	15.1	18.3	27.9	30.4	29.1
	Cr	50	22	31	30.2	21.5	25.1	23.9	24.6	24.3	32.3	13	18.5	37.4	11.6	17.7
	Ma	50	75	60	46.5	63.4	53.7	42.8	48.6	45.5	47.3	19.8	27.9	50.6	15.1	23.2
	Mi	46	31	43	37.1	35.1	36.0	34.6	23.1	27.7	26.4	27.1	26.8	24	39.6	29.9
	Tr	38	27	32	30.9	22.6	26.1	28.9	16.3	20.9	17.2	75.1	28	17	60.3	26.6
Mozilla	Bl	80	51	62	68.1	54.4	60.5	45.8	59	51.5	36.6	58.7	45.1	51.8	60.1	55.6
	Cr	55	47	51	57.3	44.3	49.9	51.1	51.6	51.3	61.2	34.8	44.4	71.6	43.7	54.3
	Ma	41	58	48	38.0	45.0	41.2	35.8	36.4	36.1	48	34.3	40	45.4	33.8	38.7
	Mi	57	41	48	43.3	47.3	45.2	37.3	38	37.6	39.1	28.6	33.1	41.9	54.8	47.5
	Tr	35	42	38	36.3	34.7	35.5	30.6	23	26.3	29.5	58.3	39.2	34.3	43.5	38.4

CNN in classifying the Trivial class. It is noteworthy that the previous models: NB [8], NBM [7] and KNN [8] utilized the BOW model in representing the bug reports. As could be observed, utilizing the embedding vectors to train the KNN and NB boost their classification performance (Table 10.3). Although the model proposed in [15] utilized the topic modeling to capture the semantic similarity between the bug reports but its performance is less than the CNN model across the five severity classes.

10.5 Related Work

This section introduces previous research in automating the identification of the bug severity level. An early research paper in this field by Menzies and Marcus [5] proposed a model for severity level identification that is based on a rule learning methodology and the information gain feature selection method. Lamkanfi et al. [6] utilized Naïve Bayes (NB) to classify the bugs as severe or non-severe. Lamkanfi et al. [7] extended their work in [6] by comparing among four machine learning techniques: NB, Naïve Bayes Multinomial (NBM), 1-nearest neighbor and Support vector machine (SVM). Experimental results over Eclipse and Gnome repositories demonstrated the superiority of the NBM in classifying the bugs into severe and non-severe. Chaturvedi and Singh [8] also compared the accuracy of five machine learning techniques: NB, K-NN, NBM, SVM, J48 and RIPPER in assigning one of four severity levels to the bug reports of the NASA dataset. The experimental results demonstrated that K-NN can achieve the highest accuracy when the value of K is tuned. Sharma et al. [10] also showed that K-NN is superior to the NBM. Yang et al. [9] investigated the impact of three feature selection methods on the performance of the NBM in classifying the bugs into severe and non-severe. Hamdy and Ellaithy [14] proposed utilizing an adapted K-NN, that is based on a distance weighted voting scheme instead of the standard majority scheme. They provided a study to its performance under different values of K, different values to the input features. It is noteworthy, that all the studies mentioned above are based on the BOW model. Roy and Rossi [11] studied the impact of including the bigram frequencies into the BOW model on the accuracy of the NB. Their finding was that, the impact of the bigram on the performance is a repository dependent. Tian et al. [12] used KNN and extended BM25 (REP) textual similarity [31] to predict the five severity levels of Bugzilla. The input features to KNN are textual and categorical; textual features are unigrams and bigrams frequencies, while the categorical includes the component and the operating system. The work of Tian et al. [12] was further extended by Zhang et al. [13] who included topics proportions as features. Hamdy and Ellaithy [15] utilized the topic proportion features to cluster the historical and new bug reports. Then a dual weighted K-NN approach was used to retrieve the neighbors of the newly submitted bug from its cluster. The dual weighted K-NN assign the severity level of the new bug using a dual distance-rank weighted voting scheme.

10.6 Conclusion

In this chapter we utilized five deep learning architectures to mine open-source software bug repositories, for the purpose of predicting the severity levels of newly submitted bug reports. The five deep learning models are the CNN, two variants of the RNN (LSTM and GRU) and the hybrid CNN-LSTM and CNN-GRU architectures. It was found that the CNN architecture is the superior in this context. While the LSTM achieved the lowest performance, in spite of its ability to learn from sequential data, which makes it ideal for text classification-based problems. Although, the hybrid model CNN-LSTM could achieve a better performance than the LSTM, but its weighted macro-average F-measure is 1% less than the CNN. Furthermore, experimental results showed that utilizing embedding models to represent bug reports can boost the classification performance of the traditional machine learning models to be comparable to the deep learning models.

This work could be extended to enhance the achieved classification performance through applying one of the oversampling techniques to the bug repositories before training the deep learning models. Besides, the dense layer could be replaced with one of the classifiers that works effectively with unbalanced datasets, e.g., Naïve Bayes Multinomial and Compliment Naïve Bayes Multinomial.

References

1. G. Yang, T. Zhang, B. Lee, Towards semi-automatic bug triage and severity prediction based on topic model and multi-feature of bug reports. In Proceedings of the IEEE 38th Annual Computer Software and Applications Conference, COMPSAC'14, 2014, pp. 97–106.
2. J. Xuan, H. Jiang, Y. Hu, Z. Ren, W. Zou, Z. Luo, X. Wu, Towards Effective Bug Triage with Software Data Reduction Techniques, IEEE Transactions on Knowledge and Data Engineering, January 2015.
3. J. Uddin, R. Ghazali1, M. Mat Deris, R. Naseem, H. Shah, A Survey on Bug Prioritization, Artificial Intelligence Review, Springer, 2017, pp.145–180.
4. X. Xia, D. Lo, M. Wen, E. Shihab, B. Zhou, An empirical study of bug report field reassignment. In the Proceedings of the 2014 Software Evolution Week-IEEE Conference on Software Maintenance, Reengineering and Reverse Engineering, CSMR-WCRE'14, 2014, pp. 174–183.
5. T. Menzies, A. Marcus, Automated severity assessment of software defect reports. In the Proceeding of IEEE International Conference on Software Maintenance ICSM 2008, Sept 2008, pp. 346–355.
6. A. Lamkanfi, S. Demeyer, E. Giger, B. Goethals, Predicting the severity of a reported bug. In the Proceedings of the 7th IEEE Working Conference on Mining Software Repositories, MSR'10, 2010, pp. 1–10.
7. A. Lamkanfi, S. Demeyer, Q. D. Soetens, T. Verdonck, Comparing mining algorithms for predicting the severity of a reported bug. In the proceedings of 15th European Conference on Software Maintenance and Reengineering (CSMR), 2011, pp. 249–258.
8. K. Chaturvedi, V. Singh, Determining bug severity using machine learning techniques. In the proceedings of Sixth International Conference on Software Engineering (CONSEG), 2012, pp. 1–6.
9. C. Z. Yang, C. C. Hou, W. C. Kao, I.-X. Chen, An empirical study on improving severity prediction of defect reports using feature selection. In the Proceedings of the 19th Asia-Pacific Software Engineering Conference, APSEC'12, 2012, pp. 240–249.
10. G. Sharma, S. Sharma, S. Gujral, A novel way of assessing software bug severity using dictionary of critical terms. In the Proceedings of 4th International Conference on Eco-friendly Computing and Communication Systems, ICECCS, 2015, Procedia Computer Science, 2015, pp. 632–639.
11. N. K. S. Roy, B. Rossi, Towards an Improvement of Bug Severity Classification, 40th EUROMICRO Conference on Software Engineering and Advanced Applications, Verona, Italy, 2014.

12. Y. Tian, D. Lo, C. Sun, Information retrieval based nearest neighbor classification for fine-grained bug severity prediction. In the proceedings of 19th Working Conference on Reverse Engineering (WCRE), 2012, pp. 215–224.

13. T. Zhanga, J. Chen, G. Yang, B. Lee c, X. Luo, Towards More Accurate Severity Prediction and Fixer Recommendation of Software Bugs, The Journal of Systems and Software, 2016, pp. 166–184.

14. A. Hamdy, A. El-laithy, Using Smote and Feature Reduction for More Effective Bug Severity Prediction, International Journal of Software Engineering and Knowledge Engineering, 29(6), 2019, pp. 897–919.

15. A. Hamdy, A. El-Laithy, Semantic categorization of software bug repositories for severity assignment automation, integrating research and practice in software engineering, Integrating Research and Practice in Software Engineering, Studies in Computational Intelligence, vol. 851. Springer, 2020, pp. 15–30.

16. M. N. Pushpalatha, M. Mrunalini, Predicting the Severity of Open Source Bug Reports Using Unsupervised and Supervised Techniques, International Journal of Open Source Software and Processes, 10(1), 2019, pp. 1–15.

17. S. Guo, R. Chen, H. Li, T. Zhang, Y. Liu, Identify Severity Bug Report with Distribution Imbalance by CR-SMOTE and ELM, International Journal of Software Engineering and Knowledge Engineering, 29(2), 2019, pp. 139–175.

18. A. Hotho, A. Nurnberger, G. Paas, A Brief Survey of Text Mining, Journal for Computational Linguistics and Language Technology, 2005, pp. 19–62.

19. A. Hamdy, M. Elsayed, Automatic Recommendation of Software Design Patterns: Text Retrieval Approach, Journal of Software, 13(4), 2018, pp. 260–268.

20. A. Hamdy, M. Elsayed, Towards More Accurate Automatic Recommendation of Software Design Patterns, Journal of Theoretical and Applied Information Technology, 96(15), 2018, pp. 5069–5079.

21. A. Hamdy, M. Elsayed, Topic modelling for automatic selection of software design patterns. In Proceedings of the International Conference on Geoinformatics and Data Analysis ICGDA '18, Prague, Czech Republic, April 20th - 22nd, 2018, pp. 41–46.

22. T. Mikolov, I. Sutskever, K. Chen, G. S. Corrado, and J. Dean, Distributed Representations of Words and Phrases and Their Compositionality, Advances in Neural Information Processing Systems, 2013, pp. 3111–3119.

23. S. Hochreiter, J. Schmidhuber, Long Short-term Memory. Neural Computation, 9(8), 1997, pp. 1735–1780.

24. M. Sundermeyer, R. Schl¨uter, H. Ney, LSTM neural networks for language modeling, In Thirteenth Annual Conference of the International Speech Communication Association, 2012.

25. J. Chung, C. Gulcehre, K. Cho, Y. Bengio, Empirical Evaluation of Gated Recurrent Neural Networks on Sequence Modeling, arXiv preprint arXiv, 1412, 2014, 3555.

26. K. Cho, B. Van Merri¨enboer, D. Bahdanau, Y. Bengio, On the Properties of Neural Machine Translation: Encoder-decoder Approaches, arXiv preprint arXiv, 1409, 2014, 1259.

27. C. Baziotis, N. Pelekis, C. Doulkeridis, Deep LSTM with attention for message-level and topic-based sentiment analysis. In Proceedings of the 11th International Workshop on Semantic Evaluation (SemEval-2017), 2017, pp. 747–754.

28. H. Wei, M. Li, Supervised Deep Features for Software Functional Clone Detection by Exploiting Lexical and Syntactical Information in Source Code, In IJCAI, 2017, pp. 3034–3040.

29. T. Wen, M. Gasic, N. Mrkˇsiˊc, P. Su, D. Vandyke, S. Young, Semantically conditioned LSTM-based natural language generation for spoken dialogue systems. In Proceedings of the 2015 Conference on Empirical Methods in Natural Language Processing, 2015, pp. 1711–1721.

30. A. Hamdy, M. Tazy, Deep Hybrid Features for Code Smells Detection, Journal of Theoretical and Applied Information Technology, 98(14), 2020.

31. C. Szegedy, W. Liu, Y. Jia, P. Sermanet, S. Reed, D. Anguelov, D. Erhan, V. Vanhoucke, A. Rabinovich, Going deeper with convolutions. In Proceedings of the IEEE Conference on Computer Vision and Pattern Recognition, 2015.

32. Y. Kim, Convolutional neural networks for sentence classification. In the Proceedings of the 2014 Conference on Empirical Methods in Natural Language Processing (EMNLP), Qatar, Doha, Oct. 2014.

33. R. Johnson, T. Zhang, Semi-supervised Convolutional Neural Networks for Text Categorization via Region Embedding, Advances in Neural Information Processing Systems, 2015, pp. 919–927.

34. J. Turian, L. Ratinov, Y. Bengio, Word representations: A simple and general method for semi-supervised learning. In Proceedings of the 48th Annual Meeting of the Association for Computational Linguistics, 2010, pp. 384–394.

35. M. Pagliardini, P. Gupta, M. Jaggi, Unsupervised Learning of Sentence Embeddings Using Compositional n-gram Features, arXiv, 1703.02507, 2017.

36. H. Palangi, L. Deng, Y. Shen, J. Gao, X. He, J. Chen, X. Song, R. K. Ward. Deep Sentence Embedding Using the Long Short Term Memory Network: Analysis and Application to Information Retrieval, CoRR, abs/1502.06922, 2015.

37. Q. Le, T. Mikolov, Distributed representations of sentences and documents. In the Proceedings of the 31st International Conference on Machine Learning, Beijing, China, 2014.

38. I. Goodfellow, Y. Bengio, A. Courville, Y. Bengio, Deep Learning, Vol. 1. MIT press Cambridge, 2016.

39. NLTK: https://www.nltk.org

40. GENSIM: https://pypi.org/project/gensim/

41. S. Ioffe, C. Szegedy, Batch normalization: Accelerating deep network training by reducing internal covariate shift. In Proceedings of the 32nd International Conference on International Conference on Machine Learning. Vol. 37. JMLR, 2015, pp. 448–456.

42. N. Srivastava, G. Hinton, A. Krizhevsky, I. Sutskever, R. Salakhutdinov, Dropout: A Simple Way to Prevent Neural Networks from Overfitting, The Journal of Machine Learning Research, 15(1), 2014, pp. 1929–1958.

43. Keras: https://keras.io/

44. Scikit: https://scikit-learn.org

Chapter 11

Predict Who: An Intelligent Game Using NLP and Knowledge Graph Model

Tameem Ahmad[a], Nesar Ahmad[a],
Mohammad Saqib[a],
and Abu Hozaifa Khan[a]

*Department of Computer Engineering, Z. H. College of
Engineering & Technology, Faculty of Engineering and
Technology, Aligarh Muslim University, Aligarh, India*

Contents

11.1   Introduction 180
11.2   Literature Survey 181
11.3   Proposed System..182
11.4   Implementation and Working 184
      11.4.1   Database Organization 186
            11.4.1.1   Image Collection.......................................186
            11.4.1.2   Tokenization and POS Tagging 186
            11.4.1.3   Popularity Score Collection.....................187
      11.4.2   Working 188
11.5   Results Formation 193
11.6   Conclusion 194
Acknowledgment 195
References 195

179

11.1 Introduction

This chapter is inspired from a famous Question Answering game, "Twenty Questions". This spoken parlor game involves at least two players and is based on deductive reasoning and creativity. One of the players is selected to be the answerer and has to think of an object/subject, but he doesn't have to reveal it to other players. All the remaining players ask questions turn by turn. Each question can be answered in "Yes" or "No". In other variants of this game "Maybe" or "Don't Know" can also be used as an answer. The answerer has to answer every question asked by all the rest of the players turn by turn. Questions could be like: "Is it white in color?" or "Is it edible?" Each question needs to be answered correctly, lying or wrong information is not allowed. If a player, asking question, guesses the answer correctly that player wins the game and now can become the answerer for the next round. If the answer isn't guessed in twenty questions then the answerer wins the game and may continue for the next round as an answerer.

The probability of winning the game for the player asking question depends on the question he/she is asking; a clever question may increase the chances for his/her winning. For example, a question such as "It runs on electricity?" can help the player to narrow down a large range of objects that require electricity to function when answered "Yes" or "No".

The domain of possible answer can be reduced by a stream of successive questions based on the reply of the answerer. Further the reply "no" reduces the possibility of a large number of objects [1]. The questions proceed with reducing the number of possible answers and finally returning a single entity.

The development of this application explores Natural Language Processing (NLP) capabilities and strives for the amalgamation of Artificial Intelligence (AI).

NLP can be defined as a branch of computer science and AI that deals with the interactions between human or natural languages and the machine such as computer [2]. AI is the branch of computer science, which helps in the creation of intelligent machines or systems that work and behave like human beings [3, 4].

A question answering (QA) system [5–7], a field of information retrieval and NLP are the systems that help the questioner to automatically find precise and concise answers to any arbitrary questions framed in human or natural language. A QA system can provide the information requested by the user [8], unlike the search engines [9, 10] such as Google or Yahoo, which only refer to the documents that are available on the web. Consider an example, suppose the question being asked "Who is the first Prime Minister of India?", ideally a QA system would respond with the answer "Jawaharlal Nehru".

An example of QA system is IBM Watson. This QA system performs a detailed analysis of the question being asked by the user to determine what is being asked and how effectively the system should approach to find the best answer. The analysis of the question involves Watson's parsing and semantic analysis technique: a deep Slot Grammar parser, a named entity recognizer, a co-reference resolution component and a relation extraction component [11].

Another example is START developed by InfoLab Group at the MIT Computer Science and Artificial Intelligence Laboratory [12, 13]. START is a QA system that responds the users with "just the right information", instead of displaying a list of information that are available on the web. Currently, the START QA system can provide answer to millions of questions in English such as about places that include cities, lakes, countries, weather, coordinates, maps, demographics, political and economic systems. It can also provide information about movies such as its titles, directors, actresses, actors, etc. and it can also provide information about personalities such as date of birth, biographies, etc.

A general QA system such as START, IBM Watson [11] and many others perform detailed analysis of the question asked by the user and approaches to retrieve the best possible answer whereas this proposed "Predict Who" asks a series of questions to the user and based on the answer given by the user (answerer), the system tries to reveal the potential character.

So far in the introduction that discusses the inspiration, motivation and the background of the work followed by Literature Survey which consists of detailed analysis of NLP and Knowledge Graph which forms the base for the project. The rest of the paper is organized in six sections. Section 11.2 is the overview of the literature that discusses a few basic underling concepts that relate to this work. Section 11.3 presents the proposed system and discusses the underlying parts of the proposed system. Section 11.4 provides the implementation details. It covers all the development tools and programming logic required for the development of the project and then discusses its working and shows the steps in details. Sections 11.5 and 11.6 summarize results and conclusion respectively.

11.2 Literature Survey

Natural language processing: NLP is the technique by which a computer can identify, recognize and understand spoken human speech. NLP makes it possible for an AI program to receive conversational input, break syntax down to comprehend the input's meaning, determine appropriate action and respond in a colloquial manner. The development of NLP applications is quite a difficult task because earlier, computers required humans to "speak" through a limited or fixed number of clearly enunciated voice or speech commands or by using a computer programming language that is accurate and more precise, unambiguous and highly structured. Human speech is generally not precise due to ambiguity, and the structure of the language used depends on various complex variables such as slang, dialects spoken in a regional language and with reference to social context [15].

Knowledge graph: It is a representational scheme [16]—in other word a "graph"—it depicts the real world entities with one another: things, not strings. It is the initial step toward searching process, which comprises the intelligence of the data on the web and understands the world just like we humans do [17]. The knowledge graph of Google currently has more than 500 million objects, 3.5 billion facts and relationships existing between these different objects [18]. And it is generally based upon what is searched on the internet and what is given as output for our search from the web [17, 19].

Decision tree: Decision trees are like binary trees. It consists of a root node, leaf nodes and branches [20–22]. The decision tree starts from the root node; both root node and every leaf node carry a question or criteria whose answer is used to traverse the tree accordingly. Branches form the link which connects a child node with its parent except for root node (a root node has no parent). Every node has at least two nodes, or it can have more than two nodes originating from it. Suppose, if the node carrying a question requires binary answer say, "yes" or "no", then there will be a child node for "yes" and another child node for "no".

One of the existing implementation of this type of system is Akinator. Akinator is a web-based game and mobile app [23], which is derived from the same popular game named as Twenty Questions that tries to determine which character the questioner is thinking of. It asks the player a series of questions. Akinator is an AI based program that can frame and find appropriate questions to ask the player. This web-based game became popular in Europe in 2009, occupying the top position on Play store and Apple store. The cartoon genie asks the question to the player while playing the game. In order to start with the game, the player must think of a famous character belonging to the field such as music, athletics, politics, cineworld, popular TV series or drama character, Internet personality, etc. The game then starts to ask a series of questions as many as required with "Yes", "No", "Probably", "Probably not" and "Don't know" as possible answers, in order to predict the possible character. If the game is able to reach to a single character within twenty-five questions then it asks the user to check if the character guessed is correct or not. If the character predicted isn't right and the game fails in three successive attempts then the game will ask the user to input the correct answer in order to expand its database and improve itself. This game makes use of tree match algorithm to predict the answer. The beauty of the game is that it reduces the number of questions required to reach the result the next time the user would think the same character. However, the implementation details and the approach haven't been revealed in the literature. Further the database is very concise and can produce answers for very few worlds' famous personalities.

This work proposes a scalable robust system with AI capability and an optimization approach to reduce the number of questions the system would require to deduce the results.

11.3 Proposed System

The proposed system is depicted in Figure 11.1. For developing this application, crawling and scraping [2, 35] of web pages are done to gather information. This collected raw data is organized as ordered dataset [24], using the concept of knowledge graph. Various Python modules have been used for collecting information, PHP and JavaScript for managing exchange of information between client and server concurrently [25]. NLP features such as parts of speech tagging and tokenization [15] are used for collection of user-specific information [26].

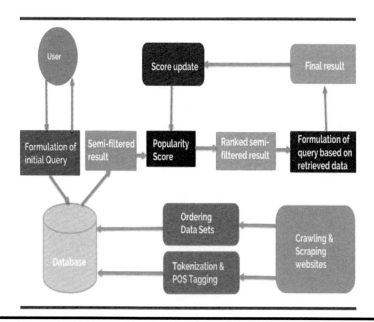

Figure 11.1   Proposed system; "Predict who".

1. *Crawling and Scraping*: This module is used to collect information for the dataset. It includes scraping websites that can provide the appropriate information about entities from various domains.
2. *Ordering datasets*: This module organizes the information gathered from crawling and websites to form ordered datasets.
3. *Tokenization & POS tagging*: The HTML paragraphs scraped from the websites are broken down into sentences, and each sentence is then processed by tokenization followed by POS tagging to extract the useful information.
4. *Database*: It holds the information of the available entities in the form of ordered datasets.
5. *Formulation of initial query*: As the user initiates its interaction with the application, the application presents some initial questions common to all entities to the user, which helps in the formulation of a basic query based on common attributes.
6. *Semi-filtered result*: After the execution on initial query on the database, the database supplies a set of entities partially filtered according to the initial query.
7. *Popularity score*: This is the value assigned to each character depending upon its listing in the Google's search result of a category.
8. *Ranked semi-filtered result*: The semi-filtered result is ranked based on popularity score, the character with highest score is ranked first.
9. *Formulation of query based on retrieved result*: Based on ranked semi-filtered characters questions are asked using characters' link, past or present attributes.

10. *Result*: When a user answers "Yes" to a specific question then character linked with that question is presented as the character predicted by the application. Otherwise the user hasn't answered all the questions correctly or the character is not present in the database.

11. *Score update*: After the application has predicted the character based on the questions answered by the user, then the popularity score of that character is updated to increase its popularity.

11.4 Implementation and Working

The working of the system is depicted in the following pseudocode that describes the complete working of the system from data collection to program execution at run-time.

Algorithm 11.1 Pseudocode in Three Parts That Depicts the Complete System Working: Part 1

```
Data Collection Procedures
Procedure assignScore:
      Assign score to entities in the order of their
appearance in Google search results
End procedure
Procedure fetchList(keywords):
      Fetch Google search URL with keywords
      Crawl Google search results to fetch list of entities
      Call assignScore
End procedure
Procedure fetchInfo (listName):
      Open listName.csv CSV file
      For each row do
              Fetch Google search URL with row->name
              Crawl Google search results to fetch suitable
                web resource
              Extract common attributes from the web
                resource
              Fetch Wikipedia page of row->name
              Apply tokenization and POS Tagging to extract
                specific attributes
              Save info in bios.csv
      End for
End procedure
Procedure fetchImages:
      Open bios.csv
      For each row do
              Fetch Wikipedia page of row->name
              Download and save image in directory images/
      End for
End procedure
```

Algorithm 11.2   Pseudocode in Three Parts That Depicts the Complete System Working: Part 2

```
Data Collection Steps
Call procedure fetchList
Call procedure fetchInfo
Call procedure fetchImages
```

Algorithm 11.3   Pseudocode in Three Parts That Depicts the Complete System Working: Part 3

```
Working of Application
Client Side:
Procedure markResponse(attribute,response):
        Set current attribute value with response
        If common attributes complete do
                Call procedure
askSpecificQuestion
        End if
        Else do
        Call askNextQuestion
End procedure

Procedure askNextQuestion(serverResponse):
        Parse question JSON from serverResponse
End procedure

Procedure askSpecificQuestion(serverResponse):
        Parse question JSON from serverResponse
End procedure

Client Side Steps:
Initialise character JSON with null values
Call procedure markResponse

Server Side Steps
Parse client request
If any unset attribute left or entities count > 4 do
For each unset attribute do
                Find attribute A with maximum
entities
        End for
        Send A as next question to client
End if
Else do
        Sort remaining entities in decreasing order
of their score
        Send list to client
End
```

11.4.1 Database Organization

Common Attributes Collection: Code in Figure 11.2.

Scraped HTML is parsed by looking for keywords such as "Profession", "Nationality", "Religions", etc. [27]. For other attributes such as gender, if while parsing "Husband" keyword is found, then the character is most likely to be a female and its gender attribute is set to 0 (representation used for gender attribute for females) otherwise the character is male and it's gender attribute is set to 1 (representation used for gender attribute for males).

In either of the cases the character is married, and its married attribute is set to 1 (representation used for married attribute for married characters) otherwise the character is unmarried and its married attribute is set to 0 (representation used for married attribute for unmarried characters).

11.4.1.1 Image Collection

Image for characters (shown in Figure 11.3) are collected from their respective Wikipedia pages and are stored by their names inside images directory.

11.4.1.2 Tokenization and POS Tagging

Paragraphs parsed from HTML collected from Wikipedia or other web resources are provided to this script. These paragraphs are first split into sentences, and each

```
try:
    rowHtml = html.find(attrs={'class':'row-'+str(i)}).text.encode("utf-8")
    rowHtml = rowHtml.replace('\n',' ')
    if 'Real Name' in rowHtml:
        data['name'] = rowHtml[10:]
    elif 'Profession' in rowHtml:
        data['profession'] = rowHtml[11:]
    elif 'Eye Colour' in rowHtml:
        data['eye_color'] = rowHtml[11:]
    elif 'Hair Colour' in rowHtml:
        data['hair_color'] = rowHtml[12:]
    elif 'Nationality' in rowHtml:
        data['nationality'] = rowHtml[12:]
    elif 'Religion' in rowHtml:
        data['religion'] = rowHtml[9:]
    elif 'Death' in rowHtml:
        data['alive'] = 0
    elif 'Husband' in rowHtml:
        data['gender'] = 0
        if not 'N/A' in rowHtml:
            data['married'] = 1
    elif 'Wife' in rowHtml:
        data['gender'] = 1
        if not 'N/A' in rowHtml:
            data['married'] = 1

    print rowHtml

except Exception as e:
    print e
```

Figure 11.2   Sample screenshot for attribute retrieval.

```
try:

    imageHtml = html.findAll(attrs={'class':'image'})
    children = imageHtml[0].find_all('img')
    picName = string.replace(id,' ','_')
    picName = string.replace(id,'.','')
    picName = string.replace(id,'"','')
    picName = picName.lower()
    url = children[0]['src']
    print url
    urllib.urlretrieve("http:"+url,'images/'+picName+'.jpg')
except Exception as e:
    print e
return
```

Figure 11.3   Sample screenshot to store relevant images of the personalities.

sentence is then analyzed using NLP to extract specific information. This script uses word tokenize and pos_tag functions respectively from NLTK to tokenize each sentence and to tag parts of speech such as Noun, Adjectives to extract useful specific information (shown in Figure 11.4).

11.4.1.3 Popularity Score Collection

Popularity score assignment is shown in Figure 11.5. For initializing popularity score, a category is provided to the Google search engine to get list of characters of that category. For a heuristic score [4], it is assumed that the order of characters appearing in the Google's search result beginning from the most popular character of that category in the search results, and then characters are scored according to these results [28].

With these components, the database of the system organized [29] in a form of a knowledge graph having extended details of different dimensions of about 305

```
from nltk import pos_tag, word_tokenize

try:

    paragraph = "His tumultuous relationship with Aishwarya Rai, his hunting of endangered species,
    sentences = paragraph.split('.')
    for sentence in sentences:
        tokenizedWords = word_tokenize(sentence)
        posTagged = pos_tag(tokenizedWords)
        print posTagged

except Exception as e:
        print e
```

Figure 11.4   Tokenization and POS tagging.

```
from selenium.webdriver.common.keys import Keys
driver = webdriver.Chrome('C:\Users\MSaqib\Downloads\Compressed\chromedriver_win32\chromedriver.exe')
tag = "famous+wrestlers"
driver.get("https://google.co.in/search?q="+tag)

cards = driver.find_elements_by_xpath("//div[@class='kltat']")
```

Figure 11.5   Sample screenshot for popularity score collection.

famous personalities of the world of all time in different categories, such as actors, cricketers, footballers, politicians, industrialists, scientists, social activists, etc.

11.4.2 Working

The principle of this application is to ask a user a number of questions relating to some entity/personality [30, 36], and then find out what is this entity. The core of the algorithm is to find the "most useful question" at each round [37, 38].

The "most useful question" being defined as the question that gives the most relevant information [31], in the optimal case splitting the audience of candidate entities into two equal halves.

The level of question starts from questions common to all entities and then slowly growing to an entity-specific question, resulting in most approximate entity. A glimpse of the way to narrow down the domain is shown in Figure 11.6.

The first step from where the game will start is shown in Figure 11.7.

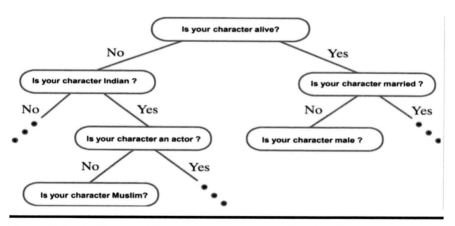

Figure 11.6   A glimpse of the way to narrow down the domain.

Figure 11.7   First screen to start the game.

Suppose the user has thought of "Mother Teresa" (present in dataset), now the interaction of user with the application will proceed as: Q1: Is your character alive? ≫ A1: No ≫ Q2: Is your character married? ≫ A2: No ≫ Q3: Is your character female? ≫ A3: Yes ≫ Q4: Is your character only non-Indian to receive Bharat Ratna? ≫ A4: Yes ≫**Your Character is Mother Teresa**. Thus, according to the answers of the questions asked the query generated was able to fetch the appropriate result.

Now let's consider another example where the user has thought of "Mahendra Singh Dhoni" (present in the dataset). Now the user interaction will proceed as: Q1: Is your character alive? ≫ A1: Yes ≫ Q2: Is your character married? ≫ A2: Yes ≫ Q3: Is your character Male? ≫ A3: Yes ≫ Q4: Is your character Indian? ≫ A4: Yes ≫ Q5: Is your character Hindu? ≫ A5: Yes ≫Q6: Is your character Politician? ≫ A6: No ≫ Q7: Is your character Actor? ≫ A7: No ≫ Q8: Is your character Cricketer? ≫ A8: Yes ≫ Q9: Is your character Indian Cricket Captain? ≫ A9: No ≫ Q10: Is your character known as "Mahi"? ≫ A10: Yes ≫**Your Character is Mahendra Singh Dhoni**.

Now let's consider another example with screenshots where the user has thought of "David Beckham" (present in the dataset), now the user interaction will proceed as shown in Figures 11.8 to 11.18:

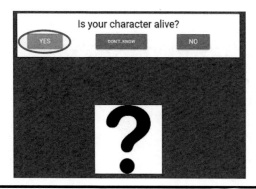

Figure 11.8   Answer: Yes.

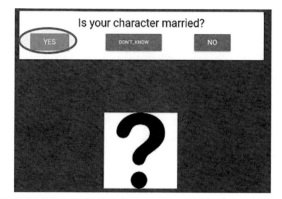

Figure 11.9   Answer: Yes.

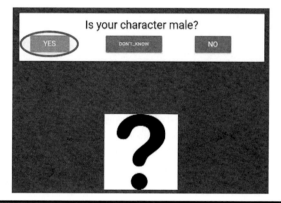

Figure 11.10   Answer: Yes.

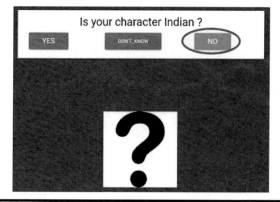

Figure 11.11   Answer: No.

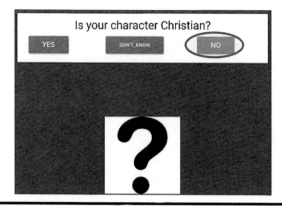

Figure 11.12   Answer: No.

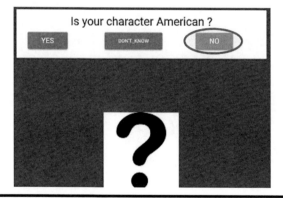

Figure 11.13   Answer: No.

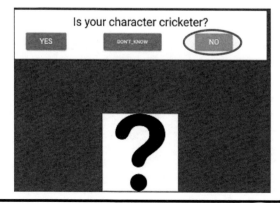

Figure 11.14   Answer: No.

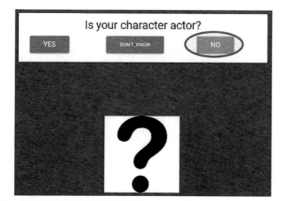

Figure 11.15   Answer: No.

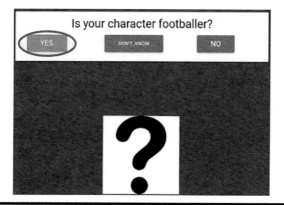

Figure 11.16   Answer: Yes.

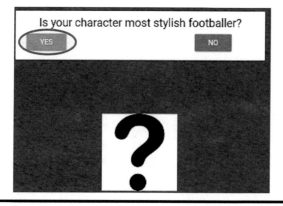

Figure 11.17   Answer: Yes.

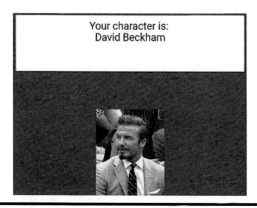

Figure 11.18   Answer: David Beckham.

Again, according to the answers of the questions asked, the query generated was able to fetch the appropriate result.

11.5 Results Formation

Final result criteria: The final result is displayed according to the following three cases:

- After filtering the dataset on every asked question to the user if only one character is left in the database, then the system returns the same character as the final answer because it would be the only solution.
- After a series of questions, if very few numbers of characters are left in the database, the heuristic search approach should be applied to predict the final answer. Every character left in the database would be slightly different from the other characters in one or other way. The heuristic approach is applied based on the pre-set score for each character. The system will try to predict the final answer based on the score in decreasing fashion, i.e., the question related to the character with highest score will be asked. This heuristic approach substantially reduces the number of questions to be asked. The demerit is that number of questions will increase in the case of less famous character.
- While predicting the answer if the total number of questions to be asked to the user is over, then either the character is not present in our database or the user has wrongly answered one or more questions of the series.

Searching process:
1. The next attribute and its value to be used in the next question is selected on the basis of largest filtering that an attribute value can provide with the currently set filters.

2. Except for binary attributes (gender, marital status, etc.) if any answer to a question is in negative then filter is applied accordingly but the attribute is still left unset, and a question will be asked again on the same attribute accordingly, until either the attribute is not needed or the users answers "Yes" to a particular value.
3. Whenever the number of characters is filtered to a reasonable number (less than four) or there are no common attributes left then the system will start to ask specific questions on the remaining characters.
4. If all characters have been reviewed and the user hasn't accepted any character, then either the character thought by the user is not in the database or the user has answered any question incorrectly.
5. If a character is accepted by the user then the score for that particular character is incremented by one to improve its popularity.

Adaptiveness:
As the system will be used over time, and the pattern of the series of questions asked and their answers are analyzed to pre predict the possible character, for which the specific question will be asked in between for the most probable character if the user accepts as its character, then the entire process will be completed in questions less than actually required otherwise the system will proceed as it is by adding an additional filter of the denied character.

11.6 Conclusion

Development of this application requires an extensive use of NLP and implementation of Natural Language Toolkit functions such as Tokenization and POS Tagging along with development of various scrappers in Python for parsing dynamic and static HTML from various web resources. And the implementation of decision tree for filtering of the characters along with the logic building of question and attribute selection and the updating of popularity score for the efficient performance of the application, and this application was successfully implemented on a dataset of 300+ characters. This application exposes to organization of ordered dataset from raw data. Crawling and scraping of web pages to gather information and organization using various Python modules, managing exchange of information between client and server concurrently. The long-term future work involves beta production of the application to users for the expansion of the database and to design and implement a more user-friendly secure [32–34] interface across various platforms.

The future work for the project will include:

1. Exploring different web resources to scrape/collect information to expand the dataset to thousands of entities and formation of ordered dataset.
2. Improving efficiency of application by minimizing the number of questions to predict the result.

3. Number of questions required can be minimized by analyzing the semi-filtered dataset using POS tagging of NLP, which will help in retrieving the remaining values.
4. Development of an automatic module to collect information of a missing character itself on providing by the user.

Acknowledgment

This work was supported by the Visvesvaraya Ph.D. Scheme for Electronics and IT fellowship of Ministry of Electronics and Information Technology (Meity), with awardee number MEITY-PHD-2979, Government of India.

References

1. "Twenty Questions: spoken parlor game," *Wikipedia, the Free Encyclopedia*. [Online]. Available: https://en.wikipedia.org/wiki/Twenty_Questions. [Accessed: 05-May-2020].
2. M. Schmitz, R. Bart, S. Soderland, and O. Etzioni, "Open language learning for information extraction," *Emnlp*, no. July, pp. 523–534, 2012.
3. S. J. Russell and P. Norvig, "Artificial Intelligence: A Modern Approach," 2010.
4. J. Turian, L. Ratinov, Y. Bengio, and J. Turian, "Word representations: A simple and general method for semi-supervised learning," *Proc. 48th Annu. Meet. Assoc. Comput. Linguist.*, no. July, pp. 384–394, 2010.
5. S. A. Khillare, B. A. Shelke, and C. N. Mahender, "Comparative study on question answering systems and techniques," *Int. J. Adv. Res. Comput. Sci. Softw. Eng.*, vol. 4, no. 11, pp. 775–778, 2014.
6. A. Ben Abacha and P. Zweigenbaum, "MEANS: A medical question-answering system combining NLP techniques and semantic Web technologies," *Inf. Process. Manag.*, vol. 51, no. 5, pp. 570–594, 2015.
7. B. Sneha, D. Mohit, and V. Zorawar Singh, "Comparison of Different Similarity Functions on Hindi QA System," *Springer, Singapore*, 2016, pp. 657–663.
8. W. Chen, Q. Zeng, L. Wenyin, and T. Hao, "A user reputation model for a user-interactive question answering system," *Concurr. Comput. Pract. Exp.*, vol. 19, no. 15, pp. 2091–2103, 2007.
9. U. Furbach, I. Glöckner, H. Helbig, and B. Pelzer, "LogAnswer – A deduction-based question answering system (system description)," in Lecture Notes in Computer Science (including subseries Lecture Notes in Artificial Intelligence and Lecture Notes in Bioinformatics), 2008, vol. 5195 LNAI, pp. 139–146.
10. T. Dong, I. Glöckner, and B. Pelzer, "A Natural Language Question Answering System as a Participant in Human Q & amp; A Portals," *IJCAI*, 2011.
11. A. Lally *et al.*, "Question analysis: How Watson reads a clue," *IBM J. Res. Dev.*, vol. 56, no. 3.4, pp. 2:1–2:14, 2012 doi: 10.1147/JRD.2012.2184637.
12. B. Katz, "Using english for indexing and retrieving," *RIAO*, 1988.
13. B. Katz, G. Borchardt, and S. Felshin, "Natural language annotations for question answering," *Proc. 19th Int. FLAIRS Conf.*, 2006.

14. A. Lally *et al.*, "Question analysis: How Watson reads a clue," 2012.

15. D. J. and J. H. Martin, Speech and Language Processing. 2009.

16. M. Taheriyan, C. A. Knoblock, P. Szekely, and J. L. Ambite, "A graph-based approach to learn semantic descriptions of data sources," in Lecture Notes in Computer Science (including subseries Lecture Notes in Artificial Intelligence and Lecture Notes in Bioinformatics), 2013, vol. 8218 LNCS, no. PART 1, pp. 607–623.

17. H. Lin, Y. Liu, W. Wang, Y. Yue, and Z. Lin, "Learning entity and relation embeddings for knowledge resolution," *Procedia Computer Science*, vol. 108, pp. 345–354, 2017.

18. A. Singhal, "Official Google Blog: Introducing the Knowledge Graph: things, not strings," *Off. Google Blog*, pp. 1–8, 2012.

19. K. Bollacker, C. Evans, P. Paritosh, T. Sturge, and J. Taylor, "Freebase: A collaboratively created graph database for structuring human knowledge," *SIGMOD 08 Proc. 2008 ACM SIGMOD Int. Conf. Manag. data*, pp. 1247–1250, 2008.

20. R. L. P. Chang and T. Pavlidis, "Fuzzy decision tree algorithms," *IEEE Trans. Syst. Man. Cybern.*, vol. 7, no. 1, pp. 28–35, 1977.

21. S. R. Safavian and D. Landgrebe, "A survey of decision tree classifier methodology," *IEEE Trans. Syst. Man Cybern.*, 1991.

22. S. R. Safavian and D. Landgrebe, "A survey of decision tree classifier methodology," *Electr. Eng.*, vol. 21, no. 3, pp. 660–674, 1991.

23. "Akinator," *Wikipedia, the free encyclopedia.* [Online]. Available: https://en.wikipedia.org/wiki/Akinator. [Accessed: 05-May-2020].

24. B. Min, R. Grishman, L. Wan, C. Wang, and D. Gondek, "Distant supervision for relation extraction with an incomplete knowledge base," in Proceedings of the 2013 Conference of the North American Chapter of the Association for Computational Linguistics: Human Language Technologies, 2013, no. June, pp. 777–782.

25. M. Haroon and T. Ahmad, "Server controlled mobile agent," *Int. J. Comput. Appl.*, vol. 11, no. 4, pp. 13–16, 2010.

26. N. Akhtar, H. Javed, and T. Ahmad, "Searching related scientific articles using formal concept analysis," in 2017 International Conference on Energy, Communication, Data Analytics and Soft Computing (ICECDS), 2017, pp. 2158–2163.

27. M. E. Califf and R. J. Mooney, "Relational learning of pattern-match rules for information extraction," *Comput. Linguist.*, vol. 4, pp. 9–15, 1999.

28. D. C. Gondek *et al.*, "A framework for merging and ranking of answers in DeepQA," *IBM J. Res. Dev.*, vol. 56, no. 3.4, p. 14:1–14:12, 2012.

29. T. Ahmad, S. Ahmad, and M. Jamshed, "A knowledge based Indian agriculture: With cloud ERP arrangement," in Proceedings of the 2015 International Conference on Green Computing and Internet of Things, ICGCIoT 2015, 2016.

30. V. Varshney, A. Varshney, T. Ahmad, and A. M. Khan, "Recognising personality traits using social media," in 2017 IEEE International Conference on Power, Control, Signals and Instrumentation Engineering (ICPCSI), 2017, pp. 2876–2881.

31. Zięba, M., Tomczak, J., and … K. B.-S. A. and, "Asking Right Questions Basing on Incomplete Data Using Restricted Boltzmann Machines". In: Świątek J., Borzemski L., Grzech A., and Wilimowska Z. (eds) Wrocław. Oficyna Wydawnicza Politechniki Wrocławskiej, Wrocław, 2014, pp. 23–32.

32. Somya, Bansal P., Ahmad T. "Methods and Techniques of Intrusion Detection: A Review". In: Unal A., Nayak M., Mishra D., Singh D., Joshi A. (eds) Smart Trends in Information Technology and Computer Communications. SmartCom 2016. Communications in Computer and Information Science, vol 628. Springer, Singapore, 2016.

33. Ahmad, Tameem, Anwar, Mohd Asad, and Haque, Misbahul, "Machine Learning Techniques for Intrusion Detection". Handbook of Research on Intrusion Detection Systems. IGI Global, 2020. 47–65. Web. 8 May. 2020. doi:10.4018/978-1-7998-2242-4.ch003

34. Tameem Ahmad, Sayyed Usman Ahmed, Syed Omar Ali, and Rifa Khan, "Beginning with exploring the way for rumor free social networks," *Journal of Statistics and Management Systems*, vol. 23, no. 2, pp. 231–238, 2020. doi: 10.1080/09720510.2020.1724623

35. Palak Bansal, Somya, Nazar Kamal, Shreya Govil, and Tameem Ahmad, "Extractive review summarization framework for extracted features," *International Journal of Innovative Technology and Exploring Engineering (IJITEE)*, vol. 8 no. 7C2, pp. 434–439, 2019.

36. O. Raghib, E. Sharma, T. Ahmad, and F. Alam, "Emotion analysis and speech signal processing," 2017 IEEE International Conference on Power, Control, Signals and Instrumentation Engineering (ICPCSI), Chennai, 2017, pp. 2872–2875, doi: 10.1109/ICPCSI.2017.8392246.

37. Hailong Huang, "Intelligent pathfinding algorithm in web games," The International Conference on Cyber Security Intelligence and Analytics. Springer, Cham, 2020.

38. Ying Tang, *et al.*, "A personalized learning system for parallel intelligent education," *IEEE Transactions on Computational Social Systems*, 2020.

Chapter 12

Mining Requirements and Design Documents in Software Repositories Using Natural Language Processing and Machine Learning Approaches

Ishaya Gambo[a,b], Clavers Chabi[b], Simon Yange[c], Rhoda Ikono[b], and Theresa Omodunbi[b]

[a]*University of Tartu, Estonia*
[b]*Obafemi Awolowo University, Ile-Ife, Nigeria*
[c]*Federal University of Agriculture, Makurdi, Nigeria*

Contents

12.1 Introduction 200
12.2 Statement of Problem 202
12.3 Related Works 202
12.4 Methodology 205
    12.4.1 Data Collection and Analysis 206
    12.4.2 Applying the Natural Language Processing 207
    12.4.3 Requirements and Design Specification 221
    12.4.4 Model Description 224

12.5 Computational Model...225
12.6 System Implementation and Performance Evaluation............................227
12.7 Results and Discussion...231
12.8 Conclusion...245
References 246

12.1 Introduction

During software development, activities such as requirements specification, system design, coding and implementation, system testing, validation, evaluation, and maintenance are crucial to the project's eventual completion. Depending on the nature of the software project at hand, these activities represent the steps needed to be followed to ensure a good quality product (Ruparelia, 2010). Each stage generates artifacts like source code, requirements and design documents, bugs report, mailing list, among several others. Also, there are conversations between stakeholders and software engineers (or developers); between developers and project managers that are recorded and saved in a particular space called a repository (Lindvall et al., 2001). In the repository, these data are usually unstructured, unorganized, and overwhelming for software developers (Bavota, 2016).

The unstructured nature of the data in the repository makes it challenging to mine and analyze (Thomas et al., 2014). The term "unstructured data" refers to the input that does not have a definite structure, non-semantic manifest, and difficult to manage because it has a lot of unlabeled, uncertain and rowdy data (Poncin et al., 2011; Thomas, 2011; Bavota, 2016). The unlabeled and noisy nature makes it very difficult to improve software development projects, while searching for a sequence of information in the repository. The importance attached to these data is due to their size; however, some contain unnecessary information, others are labeled, while some are not labeled. Also, the development team usually shares their output (e.g., they collaborate by sharing software, requirement and design documents and project plan) from the activities of a development procedure accessible by users, either for free or at a cost. Software, yet, is multifaceted, as is the development procedure at the back of it. This multi-dimensionality resides in the evidence that some works are required to produce good-quality source code, for example, conditions, coercion, records, inquiry.

In software engineering (SE) context, the activity and processes of mining various information in the repository entail proper analysis that will improve software development and evolutionary tasks (Thomas et al., 2014; Chen et al., 2016), thereby necessitating timely and error-free information. Mining software repository (MSR) in this context is the mechanism of interpreting data linked to software development methods (Thomas, 2012) for good quality products. Given the nature of software in terms of durability and widespread, it is not unusual to have repositories where developers share their lines of codes and other software artifacts (Kemerer and Slaughter, 1999). Besides, the engineering process of a software system is a versatile action, by which much output has to be produced and kept collectively. This is why software developers and project managers save all the activities and every other thing that occurs during the software development project in a repository. Software developers used these repositories to understand, analyze, and develop new software products of good quality that will satisfy users' needs (Haiduc et al., 2016).

The analysis of data can be realized by applying relevant algorithms to detect useful templates and criteria to develop divining models (Prakash et al., 2012). Software repository (SR) has non-structured input, like natural language text in bug reports and database, mailing file storage, essential record, source code comments, including accessory names (Yousef, 2015). In this chapter, we focused on requirements and design documents that contain a large amount of data or information describing "*what*" and "*how*" the system does/works from the beginning to the end. The chapter aims at developing a recommendation system that will help software developers produce a new product of good quality. About 80–85% of the data concerning the requirement and design documents in SRs are unstructured (Hassan, 2006, Hassan, 2008; Hassan and Xie, 2010), which gives many problems to the software developers to locate useful information or data needed.

Practical and exciting results can be gathered from mining unstructured SR, thereby granting developers' greater comprehension of their systems to increase their productivity profitably (Chen et al., 2016). Although many research focused on mining software repositories, still, not all the problems in the field have been solved. To achieve the goal of this chapter, we consider answering the following research question:

RQ1: how can the unstructured data in SR (e.g., requirements and design documents) be handled in a structured way for software developers to access timely and error-free information?

RQ2: what techniques can we use to force unstructured data into structured data accurately?

Remarkably, this chapter aims to solve the challenges, complexities and the peculiarities of data in order to have a structured SR. The structured repository usually provides an organized path for software developers to preserve data use in a development project (Menzies et al., 2006).

The rest of the chapter is arranged as follows: In Section 12.2, we describe the statement of the problem considered suitable for the research carried out. Section 12.3 presents a summary of existing and related works on mining unstructured software repositories alongside their limitations. In Section 12.4, we describe the methodology, data collection processes, application of Natural Language Processing (NLP) technique, requirements and design specifications, and conceptual view of our developed model. The computational model and algorithm are described in Section 12.5. In Section 12.6, the system implementation and performance evaluation are discussed. The results and discussion are presented in Section 12.7. Finally, we present the conclusion in Section 12.8.

12.2 Statement of Problem

The unstructured nature of data in software repositories is the reason behind the difficulties software developers face in mining and analyzing relevant and timely data for development purposes (Thomas et al., 2014). In particular, the inability to handle the complexity and the peculiarities of the unstructured nature of data has led to series of problems such as waste of time, loss of relevant information due to the unlabeled, vague and noisy nature of the data coming into the repositories (Bavota, 2016). As a major problem in software repositories, the unstructured data has played an essential part in the diminishing quality of new software products produce, which also is the reason behind the causes of delay during project development. For instance, to develop a project in 3 months, it can take less than 3 months if the data in the repository are structured in nature, but if not, it can take more than 3 months because much time will be spent in searching the necessary information.

Besides, forcing unstructured data into structured repositories has led to the loss of some vital data required during data processing. As such, developers waste time to locate useful information, or most times get results with some errors. This causes delays during the development process and also hinders a good way of software output. Therefore, it is necessary to device suitable ways to handle the complexities of the unstructured data. We restricted our research to the development of a model that will mine software repositories to discover useful hidden requirements and design documents and make a recommendation of the most matched with the user's request.

12.3 Related Works

Holistically, the literature offers numerous works on mining unstructured data in software repositories. Thomas et al. (2014) identified some of the SE tasks that can be made stronger with the aid of leveraging unstructured data. Emphasis on these SE tasks has been a focal point by many researchers (Canfora and Cerulo, 2005; Kiefer et al., 2007; Valetto et al., 2007; Rodriguez et al., 2012; Panichella et al., 2013; Agrawal et al., 2018). For instance, when generating requirements and design documents (Xie and Pei, 2006; Zhong et al., 2009; Wang et al., 2013; Moreno et al., 2014; Moreno et al., 2015; Abebe et al., 2016), summarizing bug reports (Rastkar et al., 2010; Mani et al., 2012; Lotufo et al., 2015) and changes in code (Buse and Weimer, 2010; Rastkar and Murphy, 2013; Cortés-Coy et al., 2014).

These works and many others have been able to tackle the complex nature of data and their peculiarity. Thus, mining unstructured SR is considered a growing and popular area of research in the SE community (Bavota, 2016).

Technically, emphasis on mining unstructured SR borders on the assessment and improvement of software quality (Poshyvanyk and Marcus, 2006; Marcus et al., 2008; Arnaoudova et al., 2013; Bavota et al., 2013). The principal goal is to build recommenders that provide supports for project managers in tasks relating to training and capacity building. More precisely, the advantage of this technical importance is tailored toward enhancing the coding activities of developers daily (Stylos and Myers, 2006; Goldman and Miller, 2009; Takuya and Masuhara, 2011; Cordeiro et al., 2012; Rigby and Robillard, 2013; Ponzanelli et al., 2014; Subramanian et al., 2014). Besides, it also aids the identification of duplicated bug reports (Wang et al., 2008). In the subsequent paragraphs that follow in this section, we attempted to describe in chronological order some of the related research and corresponding gaps that need to be filled to advance the course of study further.

Kiefer et al. (2007) used iSPARQL and Software Evolution Ontology to mine software repositories. The Unified Modeling Language (UML) tool was used to specify the design and FAMIX-based software ontology model was used for the implementation. The mining was done subject to the semantic web query SPARQL. The iSPARQL frame and EvoOnt enabled mining represented in the OWL input format. The restriction of the approach is the insufficiency of data because of the utilization of the FAMIX-based software ontology model and the complexity of the data, which makes the system's result not satisfactory. Also, Hattori et al. (2008) focused on the impact analysis of MSR in a case study. The Apriori and DAR algorithms were compared to determine which is the best. The result of the research showed that the DAR algorithm performed better than the Apriori algorithm. However, the massive scale of the repositories, development process used and the complexity of data had a significant influence on the DAR algorithm.

An automated library recommendation was developed by Elsen et al. (2009) using association rule and collaborative filtering techniques. These techniques gave an excellent recall rate in terms of recommending libraries for software developers. However, the technique was limited in terms of a textual description of libraries, hence, not suitable when considering requirements and design documents. In the work of Zhong et al. (2009) on MAPO: Mining and API Usage Patterns recommendation. The clustering approach was utilized to analyze the application program interface (API) patterns for discovering the similarities that relate to them. With this technique, MAPO mined frequent call sequences and also analyzed patterns associated with multiple API method calls. MAPO could mine patterns that are useful under programming contexts, and it requires less exercise to discover the API. However, the study conducted by the author has applied a few datasets and considered only the API. Consequently, scalability and efficiency become a significant problem.

In Thomas (2012) and Thomas et al. (2014), several tools and techniques for mining unstructured software repositories were surveyed. Emphasis was on the information retrieval (IR) model and the NLP technique for improving software evolution. The IR model and NLP technique are fast and straightforward to compute. As an effective approach, they brought useful and practical results to software development, being able to give approximately the expected output. However, the Latent semantic indexing (LSI) and Latent Dirichlet Allocation (LDA) have some incorrectly change events due to the noise in the model. This means both the IR model could not handle the noise and natural text in the repositories. Also, the ontology approach was used by Junior and Favero (2012) to detect and manage the security concerns of the repository. As such, the Sambasore approach (Junior and Favero, 2012; Kasianenko, 2018) was developed using the Java programming language. Other programming languages were not considered and supported by the Sambasore approach.

The work of Chan et al. (2012) modeled classes and methods to search connected API subgraph by using phrases. This approach allows users to enter in their phrase request and to get a linked API. However, there exist the possibility that these queries may not be the exact representation of samples to common and standard API used by others. Given that humans are involved in making these queries, it is also possible that a phrase constructed as a query may affect the results, which will eventually make the precision and sensitivity analysis to go higher or reduce. Consequently, Thung et al. (2013) developed a mechanism that automatically provides a recommendation of API methods from feature requests using a trained dataset and ranking approach.

In addition, Thung et al. (2013) approach were evaluated on change requests of five applications and recommended methods from ten libraries. The outcome of the study showed that the proposal could recommend relevant feature requests in a project with five methods of recommendation. When the recommendation size increases to ten, for example, the approach could recommend at least one suitable method for 70.9%. However, this approach was tested on a small scale dataset, and it was integrated with any IDE raison. At the same time, the performance percentage had reduced when the size increased. Progressively, Shi et al. (2015) used a feature interface graph (FIG) to recommend interfaces automatically. This was achieved with the use of LDA to extract data from open source repositories. The results of this approach made recommendations of valuable interface for programmers to reuse; however, the recommendation often comes with errors due to the size of the data and complex nature of the data.

Going further, Sun et al. (2016) conducted a study that focused on handling information that is relevant in the repository when it comes to interface recommendation. LDA was used for extracting data and identifying the right open-source repository and projects. Also, the interface is clustered using the FIG mechanism for the recommendation on the clustered result. This will typically require a short time to display the interface recommendation, which will improve efficiency. However, the limitation could be that users will not be able to set their request. Instead, they are meant to choose from the already existing request.

In summary, we observed that mining data from software repositories provide supports for project managers and developers to uncover useful insights (Buse and Zimmermann, 2012; Siddiqui and Ahmad, 2018; Gupta and Gupta, 2019a). Noticeably, the software artefacts contained in software repositories are represented in natural language form. On the one hand, the natural language form makes NLP techniques suitable for semantics and biomedical engineering research, sentiment analysis (Kharlamov et al., 2019; Polyakov et al., 2020), computational linguistics analysis and modeling (Storks et al., 2019), IR, text and multi-document summarization (Gupta and Gupta, 2019b). Besides, NLP techniques are useful for classifying software artefacts (e.g., requirements and design documents) (Nazir et al., 2017), extracting ontologies from requirements specifications to determine its completeness (Kaiya and Saeki, 2006; Arellano et al., 2015; Gupta and Gupta, 2019a; Zhong et al., 2020). On the other hand, machine learning techniques play an essential role in mining SR (Chaturvedi and Singh, 2012; Güemes-Peña et al., 2018; Siddiqui and Ahmad, 2018).

Without any doubt, numerous researches have been conducted in the field of mining software repositories such as mining software repositories to predict bugs, mailing lists, mining software repositories for source code analysis and many other artefacts in the repository. Still, we uncover some existing gaps that borders on the following fact: (i) existing techniques did not consider a large-scale dataset, (ii) existing techniques and models failed in handling the noises in natural language text, and (iii) existing techniques and models are not able to provide proper and adequate labeling of the data.

12.4 Methodology

We restricted the focus of our research to the development of a model that will mine software repositories to discover useful hidden requirements and design documents, and make a recommendation of the most matched with the user's request. We employed the quantitative research approach (Takhteyev and Hilts, 2010; Gousios et al., 2014) that provides support for measurement based on data. The quantitative approach aims at extracting data from GitHub to evaluate software development practices in the GitHub repository (Kalliamvakou et al., 2014).

Figure 12.1 shows the step-by-step mining GitHub describing the methodology. To address the complex nature of data in GitHub, we identified two types of noise that includes: (i) Natural language text comprising of special characters and English words, and (ii) Unlabeled data that are not categorized under a determined case or that are unstructured. Data is considered noisy if it contains incorrect or wrong records or outliers. However, records and observations are used in machine learning applications. Noise consists of void or null information that is not relevant

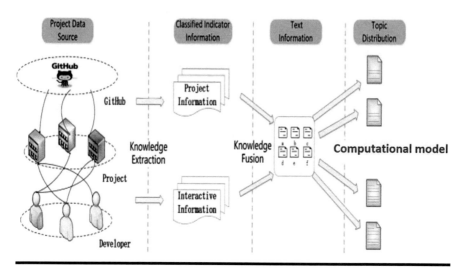

Figure 12.1   Step-by-step mining GitHub. (Adapted from Steinmacher et al., 2015.)

to the dataset. If the datasets have those errors, they become justifiable to either process or clean the data. For instance, any unlabeled data or non-specific are noise in the dataset that requires cleaning.

12.4.1 Data Collection and Analysis

The data collection source was directly from GitHub, and two methods were employed: data extraction and data processing. The data collection concerns the public repositories in GitHub for a period of 14 years (i.e., 2004–2018). One thousand (1,000) projects extracted from GitHub were used as the dataset. In particular, we used PyDriller for data collection (see Figure 12.2 for a sample of the dataset) from free repositories in GitHub. Data extraction in this context is a procedure of data recuperation from data sources (mostly non-structured or not well structured) to treat them or store them (data migration). The sequence of code in Figure 12.3 was used in PyDriller for extracting data. The dataset typically contains information like an event, author's name, source code snippet, repository link, commits, new additions, documents, communication.

The dataset was analyzed and processed using Google BigQuery and Natural Language Processing (NLP) techniques. We analyzed and processed the extracted dataset to get useful data. In particular, we analyzed the data for better understanding and to realize how these data are related to addressing the inherent complexity of the unstructured nature of the data. We regarded data to be incomplete if it lacked features or contained missing features that will not be able to capture the problem being addressed.

3561493,Is there a RegExp.escape function in Javascript?,1403,12,97,87115,javascript|regex
14967647,encode/decode image with base64 breaks image,1085,2,6,25238,javascript|fileapi
7422072,JavaScript - How To Detect Number As A Decimal (Including 1.0),975,5,3,11488,javascript
57803,How to convert decimal to hex in
JavaScript?,890,26,239,740579,javascript|hex|tostring|base|number-formatting
105034,Create GUID / UUID in JavaScript?,661,52,1210,1417654,javascript|guid|uuid
1252512,jQuery Datepicker: Prevent closing picker when clicking a
date,576,9,4,38424,javascript|jquery|jquery-ui|jquery-ui-datepicker
470832,Getting an absolute URL from a relative one. (IE6
issue),533,11,30,45105,javascript|url|internet-explorer-6
11929099,HTML5 Canvas drawImage ratio bug iOS,513,5,49,41740,javascript|ios|html5|canvas
13216903,Get binary data with XMLHttpRequest in a Firefox
extension,448,1,,1387,javascript|ajax|firefox|firefox-addon
359788,How to execute a JavaScript function when I have its name as a
string,431,32,294,444316,javascript
3143070,javascript regex iso datetime,426,6,9,42448,javascript|regex|datetime|iso
10046972,MSIE Returns status code of 1223 for Ajax Request,422,1,2,6693,javascript|ajax|internet-explorer|xmlhttprequest
43044,Algorithm to randomly generate an aesthetically-pleasing color
palette,415,16,197,124012,algorithm|colors
992461,Is it possible to override document.cookie in WebKit?,370,1,,2018,javascript|cookies|webkit
3653444,CSS styles not applied on dynamic elements in Internet Explorer
7,366,2,,6272,javascript|css|internet-explorer|internet-explorer-8|internet-explorer-7
7264899,Detect css| transitions using javascript (and without modernizr)?,338,5,13,21400,javascript-events|css3|cross-browser|css-transitions
499126,jQuery Set Cursor Position in Text Area,332,15,167,320698,javascript|jquery|html|textfield
5312849,jQuery find self,313,6,1,9819,jquery|jquery-selectors

Figure 12.2   Sample of data extracted.

```
from pydriller import RepositoryMining

for public_repo in RepositoryMining ('https://github.com/').all():
    print (public_repo)
```

Figure 12.3   Source code for extracting data.

Table 12.1 shows some of the parameters used to aid data processing, and Table 12.2 presents samples of the processed data. Usually, the data collected are not always in a usable form. For this reason, it became so important that the right data is fed into the system for the problem to be solved. In this regard, we ensured that the data was converted to a useful scale format with all the essential features extracted from the repository, and processed before putting them into the computational model. Data can be in text format or file after processing. Processing the data is an excellent way to detect abnormalities or irregularities and peculiarities within the dataset. So, the user's request and data collected from GitHub repositories were processed.

12.4.2 Applying the Natural Language Processing

To process data collected from GitHub, we use NLP by calling a function in Python to perform the task. The reason for using NLP is to allow the system to understand the meaning of the text, which improves the efficiency of the system in finding a specific piece of information from GitHub. This way, we can categorize

Table 12.1 Parameters Used for Data Extraction

SN	Elements	Description
1	Username	The user name indicates the person that created the repository
2	Repository name	The name of the repository
3	Repository description	The content of activities of the repository
4	Programming language	The programming language used for implementation in the repository
5	Number of commits, changes and stars	The total number of changes, commits and the affluence of commends
6	Repository link	This is the directory of the repository for accessing the information

the raw data extracted. Besides, we focused on the most popular repositories that programmers or developers often use to extract their data for a given project to get the required information concerning specific aspects. By this, we were able to determine the most frequently used programming language and the commits number which occur in the repository. This includes the full name of the author that added the commit or changes in the repository. Consequently, the data are also categorized based on the evolutionary changes in the repository. The goal is to handle the complex nature of data to succeed in recommending relevant results to users.

The NLP was used on the user's request and the description of the sub-repository. NLP involves steps like tokenization, splitting and stop word (Weiss et al., 2005; Bird et al., 2009; Manning et al., 2014). We used the tokenization, splitting-function and stop-words algorithms in Weiss et al. (2005) respectively. The tokenization algorithm tokenizes user requests and describes the content in the repository by putting them in sequenced words. For instance, we considered the following user's request as an example **"Django controller configuration on windows, and users backends' management":** when this goes into the tokenize Algorithm, it will return this ≪ "Django" "controller" "configuration" "on" "windows" "," "and" "users" "backends'" "management" ≫; and the same thing will happen with a repository text description. The splitting-function algorithm helps in removing non-technical words or words that are not useful. When applied to the example as mentioned earlier, the splitting function algorithm returns the following result: ≪ **"Django"** "controller" "configuration" **"windows"** "users" **"backends'"** "management" ≫. The stop-words algorithm checks for redundancy and deletes duplicated or repeated words in the sentence; in this example, there is no redundancy, so nothing changed.

Table 12.2 Sample of Processed Data

Username	Repository Name	Description	Language	Number of Stars	URL
freeCodeCamp	freeCodeCamp	The https://freeCodeCamp.com open-source codebase and curriculum. Learn to code and help nonprofits	JavaScript	290k	https://github.com/ freeCodeCamp/ freeCodeCamp
Twbs	bootstrap	The most popular HTML, CSS and JavaScript framework for developing responsive, mobile-first projects on the web	JavaScript	112k	https://github.com/twbs/ bootstrap
EbookFoundation	free-programming-books	Freely available programming books		87.8k	https://github.com/ EbookFoundation/ free-programming-books
Facebook	react	A declarative, efficient and flexible JavaScript library for building user interfaces	JavaScript	69.7k	https://github.com/ facebook/react
d3	d3	Bring data to life with SVG, Canvas and HTML	JavaScript	65.7k	https://github.com/d3/d3
Getify	You-Dont-Know-JS	A book series on JavaScript. @YDKJS on twitter	JavaScript	62k	https://github.com/getify/ You-Dont-Know-JS
Tensorflow	tensorflow	Computation using data flow graphs for scalable machine learning	C++	61.6k	https://github.com/ tensorflow/tensorflow

(Continued)

Table 12.2 (Continued) Sample of Processed Data

Username	Repository Name	Description	Language	Number of Stars	URL
Sindresorhus	awesome	A curated list of awesome lists		60.7k	https://github.com/sindresorhus/awesome
Vuejs	vue	A progressive, incrementally-adoptable JavaScript framework for building UI on the web	JavaScript	57.7k	https://github.com/vuejs/vue
Angular	angular.js	AngularJS – HTML enhanced for web apps!	JavaScript	56.2k	https://github.com/angular/angular.js
Robbyrussell	oh-my-zsh	A delightful community-driven (with 1,000+ contributors) framework for managing your zsh configuration. Includes 200+	Shell	55.2k	https://github.com/robbyrussell/oh-my-zsh
Airbnb	javascript	JavaScript Style Guide	JavaScript	54.1k	https://github.com/airbnb/javascript
Github	gitignore	A collection of useful. gitignore templates		51.6k	https://github.com/github/gitignore
FortAwesome	Font-Awesome	The iconic font and CSS toolkit	HTML	50.9k	https://github.com/FortAwesome/Font-Awesome
Facebook	react-native	A framework for building native apps with React	JavaScript	50k	https://github.com/facebook/react-native

(Continued)

Table 12.2 (Continued) Sample of Processed Data

Username	Repository Name	Description	Language	Number of Stars	URL
Electron	electron	Build cross platform desktop apps with JavaScript, HTML and CSS	C++	47.2k	https://github.com/electron/electron
Torvalds	linux	Linux kernel source tree	C	46.3k	https://github.com/torvalds/linux
Jquery	jquery	jQuery JavaScript Library	JavaScript	45.3k	https://github.com/jquery/jquery
Jwasham	coding-interview-university	A complete computer science study plan to become a software engineer		44.9k	https://github.com/jwasham/coding-interview-university
Moby	moby	Moby Project – a collaborative project for the container ecosystem to assemble container-based systems	Go	44.3k	https://github.com/moby/moby
Daneden	animate.css	A cross-browser library of CSS animations. As easy to use as an easy thing	CSS	42.8k	https://github.com/daneden/animate.css
Apple	swift	The Swift Programming Language	C++	39.1k	https://github.com/apple/swift

(Continued)

Table 12.2 (Continued)   Sample of Processed Data

Username	Repository Name	Description	Language	Number of Stars	URL
Atom	atom	The hackable text editor	CoffeeScript	37.8k	https://github.com/atom/atom
Meteor	meteor	Meteor, the JavaScript App Platform	JavaScript	37.7k	https://github.com/meteor/meteor
h5bp	html5-boilerplate	A professional front-end template for building fast, robust and adaptable web apps or sites	JavaScript	37.6k	https://github.com/h5bp/html5-boilerplate
Nodejs	node	Node.js JavaScript runtime	JavaScript	36.4k	https://github.com/nodejs/node
Nodejs	node-v0.x-archive	Moved to https://github.com/nodejs/node		36.4k	https://github.com/nodejs/node-v0.x-archive
Rails	rails	Ruby on rails	Ruby	36.1k	https://github.com/rails/rails
Semantic-Org	Semantic-UI	Semantic is a UI component framework based around useful principles from natural language	JavaScript	35.6k	https://github.com/Semantic-Org/Semantic-UI
Vinta	awesome-python	A curated list of awesome Python frameworks, libraries, software and resources	Python	35.3k	https://github.com/vinta/awesome-python

(Continued)

Table 12.2 (Continued) Sample of Processed Data

Username	Repository Name	Description	Language	Number of Stars	URL
Hakimel	reveal.js	The HTML presentation framework	JavaScript	35.2k	https://github.com/hakimel/reveal.js
Socketio	socket.io	Realtime application framework (Node.JS server)	JavaScript	33.8k	https://github.com/socketio/socket.io
Mrdoob	three.js	JavaScript 3D library	JavaScript	33.6k	https://github.com/mrdoob/three.js
Laravel	laravel	A PHP Framework For Web Artisans	PHP	32.9k	https://github.com/laravel/laravel
Expressjs	express	Fast, unopinionated, minimalist web framework for node	JavaScript	32.5k	https://github.com/expressjs/express
Reactjs	redux	Predictable state container for JavaScript apps	JavaScript	32k	https://github.com/reactjs/redux
Moment	moment	Parse, validate, manipulate and display dates in javascript	JavaScript	31.9k	https://github.com/moment/moment
Impress	impress.js	It is a presentation framework based on the power of CSS3 transforms and transitions in modern browsers and inspired by prezi.com	JavaScript	31.8k	https://github.com/impress/impress.js

(Continued)

Table 12.2 (Continued)   Sample of Processed Data

Username	Repository Name	Description	Language	Number of Stars	URL
Nwjs	nw.js	Call all Node.js modules directly from DOM/WebWorker and enable a new way of writing applications with all Web techno	C++	31.8k	https://github.com/nwjs/nw.js
Jlevy	the-art-of-command-line	Master the command line, in one page		31k	https://github.com/jlevy/the-art-of-command-line
Chartjs	Chart.js	Simple HTML5 Charts using the <canvas> tag	JavaScript	30.7k	https://github.com/chartjs/Chart.js
Google	material-design-icons	Material Design icons by Google	CSS	30.2k	https://github.com/google/material-design-icons
jakubroztocil	httpie	Modern command line HTTP client user-friendly curl alternative with intuitive UI, JSON support, syntax highlighting	Python	30.2k	https://github.com/jakubroztocil/httpie
ionic-team	ionic	Build amazing native and progressive web apps with open web technologies. One app running on everything	TypeScript	30.1k	https://github.com/ionic-team/ionic
Jekyll	jekyll	Jekyll is a blog-aware, static site generator in Ruby	Ruby	30.1k	https://github.com/jekyll/jekyll

(Continued)

Table 12.2 (Continued)   Sample of Processed Data

Username	Repository Name	Description	Language	Number of Stars	URL
Microsoft	vscode	Visual Studio Code	TypeScript	29.6k	https://github.com/Microsoft/vscode
Webpack	webpack	A bundler for javascript and friends. Packs many modules into a few bundled assets. Code Splitting allows to load par	JavaScript	29.4k	https://github.com/webpack/webpack
Resume	resume.github.com	Resumes generated using the GitHub information	JavaScript	29.4k	https://github.com/resume/resume.github.com
AFNetworking	AFNetworking	A delightful networking framework for iOS, OS X, watchOS and tvOS	Objective-C	29.4k	https://github.com/AFNetworking/AFNetworking
Golang	go	The Go programming language	Go	29.2k	https://github.com/golang/go
Facebookincubator	create-react-app	Create React apps with no build configuration	JavaScript	29.1k	https://github.com/facebookincubator/create-react-app
Homebrew	legacy-homebrew	The former home of Homebrew	Ruby	29k	https://github.com/Homebrew/legacy-homebrew
Nvbn	thefuck	Magnificent app which corrects your previous console command	Python	28.6k	https://github.com/nvbn/thefuck

(Continued)

Table 12.2 (Continued)   Sample of Processed Data

Username	Repository Name	Description	Language	Number of Stars	URL
h5bp	Front-end-Developer-Interview-Questions	A list of helpful front-end related questions you can use to interview potential candidates, test yourself or complete		28.2k	https://github.com/h5bp/Front-end-Developer-Interview-Questions
Pallets	flask	A microframework based on Werkzeug, Jinja2 and good intentions	Python	27.8k	https://github.com/pallets/flask
NARKOZ	hacker-scripts	Based on a true story	JavaScript	27.8k	https://github.com/NARKOZ/hacker-scripts
Google	material-design-lite	Material Design Components in HTML/CSS/JS	HTML	27.8k	https://github.com/google/material-design-lite
Adobe	brackets	An open-source code editor for the web, written in JavaScript, HTML and CSS	JavaScript	27.6k	https://github.com/adobe/brackets
Dogfalo	materialize	Materialize, a CSS Framework based on Material Design	JavaScript	27.2k	https://github.com/Dogfalo/materialize
Blueimp	jQuery-File-Upload	File Upload widget with multiple file selection, drag&drop support, progress bar, validation and preview images, audi	JavaScript	26.9k	https://github.com/blueimp/jQuery-File-Upload

(Continued)

Table 12.2 (Continued) Sample of Processed Data

Username	Repository Name	Description	Language	Number of Stars	URL
rg3	youtube-dl	Command-line program to download videos from YouTube.com and other video sites	Python	26.9k	https://github.com/rg3/youtube-dl
Necolas	normalize.css	A collection of HTML element and attribute style-normalizations	CSS	26.7k	https://github.com/necolas/normalize.css
Gulpjs	gulp	The streaming build system	JavaScript	26.6k	https://github.com/gulpjs/gulp
Callemall	material-ui	React Components that Implement Google's Material Design	JavaScript	26.6k	https://github.com/callemall/material-ui
Jashkenas	backbone	Give your JS App some Backbone with Models, Views, Collections and Events	JavaScript	26.4k	https://github.com/jashkenas/backbone
Django	django	The Web framework for perfectionists with deadlines	Python	26.4k	https://github.com/django/django
Yarnpkg	yarn	Fast, reliable and secure dependency management	JavaScript	25.9k	https://github.com/yarnpkg/yarn
Requests	requests	Python HTTP requests for humans	Python	25.7k	https://github.com/requests/requests

(Continued)

Table 12.2 (Continued)   Sample of Processed Data

Username	Repository Name	Description	Language	Number of Stars	URL
Zurb	foundation-sites	The most advanced responsive front-end framework in the world. Quickly create prototypes and production code for site	JavaScript	25.7k	https://github.com/zurb/foundation-sites
Angular	angular	One framework. Mobile & desktop	TypeScript	25.3k	https://github.com/angular/angular
ReactiveX	RxJava	RxJava Reactive Extensions for the JVM a library for composing asynchronous and event-based programs using observ	Java	25.2k	https://github.com/ReactiveX/RxJava
Chrislgarry	Apollo-11	Original Apollo 11 Guidance Computer (AGC) source code for the command and lunar modules	Assembly	24.4k	https://github.com/chrislgarry/Apollo-11
Kubernetes	kubernetes	Production-Grade Container Scheduling and Management	Go	24.3k	https://github.com/kubernetes/kubernetes
Lodash	lodash	A modern JavaScript utility library delivering modularity, performance & extras	JavaScript	24.3k	https://github.com/lodash/lodash
Alamofire	Alamofire	Elegant HTTP Networking in Swift	Swift	24.1k	https://github.com/Alamofire/Alamofire

(Continued)

Table 12.2 (Continued)   Sample of Processed Data

Username	Repository Name	Description	Language	Number of Stars	URL
antirez	redis	Redis is an in-memory database that persists on disk. The data model is key-value, but many different kind of values	C	24k	https://github.com/antirez/redis
Wasabeef	awesome-android-ui	A curated list of awesome Android UI/UX libraries		23.9k	https://github.com/wasabeef/awesome-android-ui
Ansible	ansible	Ansible is a radically simple IT automation platform that makes your applications and systems easier to deploy. Avoid	Python	23.9k	https://github.com/ansible/ansible
Getlantern	lantern	Lantern Latest Download https://github.com/getlantern/lantern/releases/tag/latest https://github.com/getl	Go	23.8k	https://github.com/getlantern/lantern
Josephmisiti	awesome-machine-learning	A curated list of awesome machine learning frameworks, libraries and software	Python	23.7k	https://github.com/josephmisiti/awesome-machine-learning
kamranahmedse	developer-roadmap	Roadmap to becoming a web developer in 2017		23.6k	https://github.com/kamranahmedse/developer-roadmap

(Continued)

Table 12.2 (Continued)   Sample of Processed Data

Username	Repository Name	Description	Language	Number of Stars	URL
papers-we-love	papers-we-love	Papers from the computer science community to read and discuss		23.6k	https://github.com/papers-we-love/papers-we-love
Elastic	elasticsearch	Open Source, Distributed, RESTful Search Engine	Java	23.4k	https://github.com/elastic/elasticsearch
Trinea	android-open-project	A categorized collection of Android Open Source Projects: codekk		23.3k	https://github.com/Trinea/android-open-project
Neovim	neovim	Vim-fork focused on extensibility and usability	Vim script	23.2k	https://github.com/neovim/neovim
Microsoft	TypeScript	TypeScript is a superset of JavaScript that compiles to clean JavaScript output	TypeScript	23.2k	https://github.com/Microsoft/TypeScript

12.4.3 Requirements and Design Specification

Due to the vast number of the data coming into the repositories, and the problem of the complex nature of data, we thought it necessary to provide an architectural description of our approach to addressing the existing gap in research. The architectural description, as shown in Figure 12.4, has five significant steps which include: data extraction, pattern discovery or data analysis, data mining system, categorization and interpretation. The goal for the architecture is to enable the implementation of a system that can tabulate user's request during a search, check and search for similarity between items.

Based on the architectural description, we specified the design using UML since one of our goals is to implement a prototype system. At first, we displayed the

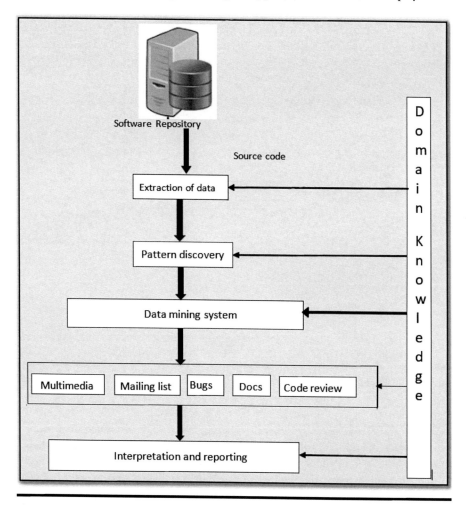

Figure 12.4 **Architecture of mining unstructured software repositories.**

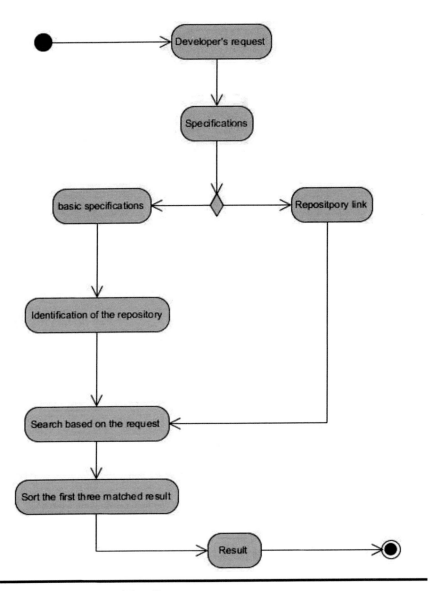

Figure 12.5 System activity diagram.

flow of activities from the start to the end, along with the decisions and conditions in the activity diagram shown in Figure 12.5. To start, developers enter in their request and the specific input. If the inputs are basic specifications, the system will compute through a particular algorithm to identify the list of the repository that matches the request and search for it. If the specification is the repository link, then the system searches for the code and brings the result out. In Figure 12.6,

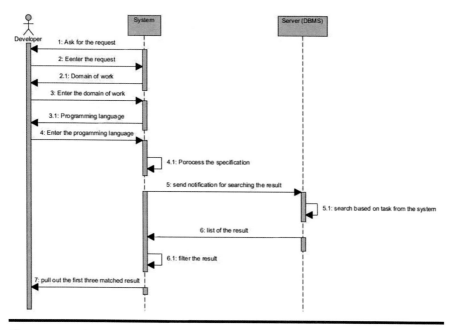

Figure 12.6 System sequence diagram.

we describe the sequence diagram where the system asks for the necessary input (the request, the domain area, the programming language, the repository link if known). After that, it collects all the data and processes it to pull out the result that the user is searching for. We also used the class diagram in Figure 12.7 to define the system's static structure by describing the classes, attributes, methods and the relationship among the objects.

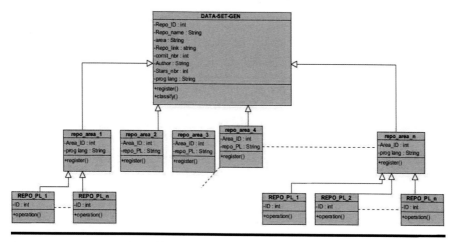

Figure 12.7 System class diagram.

12.4.4 Model Description

The model is an interactive control system which is designed in the way that stakeholders can browse easily through the clear and detailed content to get the result. Figures 12.8 and 12.9 describe the developed model and the step-by-step processes in a concepual manner, respectively. In particular, Figure 12.8 shows that the system gets the dataset from GitHub using PyDriller; the raw dataset is analyzed with BigQuery by specifying the parameters required for the analysis. The next step is to handle the noises that are in the dataset by applying NLP to the text. After this, the unsupervised learning algorithm is applied to group these texts according to their content, label and then store in a database.

The developers showed in Figure 12.8 are the end-users of the system. When a developer makes a request, the system fetches and processes such request by applying NLP on the text. The processed data is then compared with the already processed dataset to get the similarity between the request and the set of the structured data, and then finally get the matched list that will be sent to the users. In Figure 12.9, we included the parameters used and breakdown the model into four phases consisting of data extraction, processing, rating and data labeling.

In summary, the developed model collects the processed dataset and stores it on a database. The importance of processing the data is to eliminate the noise in

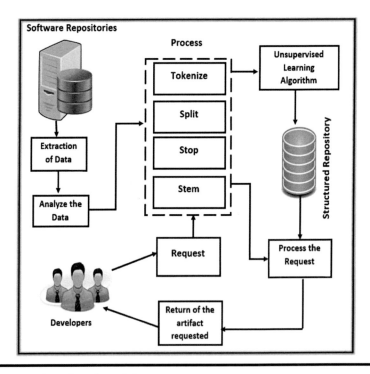

Figure 12.8 Developed model.

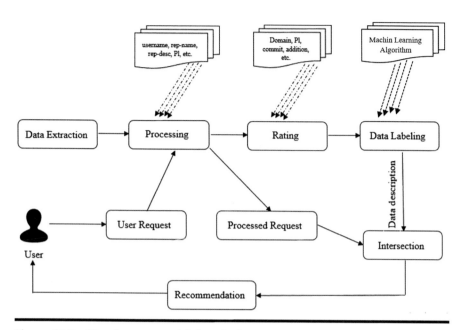

Figure 12.9   Step-by-step model description.

the data extracted or collected. The next step is to get the user request and process it in a clear way that the computer understands. By doing this, delay in accessing required data in the repository will be avoided, since the approach offers a means of making real-time recommendations of GitHub content. After this, it counts word occurrence in both user requests and repository descriptions to get the similarity between the request, and the repository that has the matched answer. Finally, we use a computational approach to determine the author's rating and the importance of commits in that repository.

12.5 Computational Model

We used the Artificial Neural Network (ANN) technique to formulate the computational model to handle unstructured data in the repositories. The outcome of the formulated model is to provide a Top-N recommendation list of sub-repositories with the necessary information containing requirements and design documents. The goal is to enable users to utilize historical interactions and attributes of the project, such as domain area, programming language and commits. The content-based methods are user commits of the project. The computational model focused on estimating the preferences of users toward items and making suggestions on items that will match their request. Thus, we assume two sets (Ur) and (I) representing users' requests and items that could be recommended to them by predicting a single rate (*r*) respectively. In this context, the single rate (*r*) serves as the degree to which a user (*u*) may be satisfied by the items recommended. The utility function is defined in Equations (12.1) and (12.2) respectively:

$$f(Ur \times i) \rightarrow r \tag{12.1}$$

$$f : Ur \text{ x } I \rightarrow R_o, R_1, \ldots\ldots R_n \tag{12.2}$$

where *n* is given as the number of criteria used and r_o the overall rating, Ro = $f(R_o, R_1, R_2, \ldots, R_n)$.

$$R_o = W_1.R_1 + W_2.R_2 + W_3.R_3 + W_4.R_4 + C \tag{12.3}$$

W represents the weight, and *C* is a constant. We have demonstrated this in Figure 12.10 depicting the conceptual view of the ANN-based System Analysis and Mining List Recommendation System (SAMLRS). We considered the criteria R_1 = domain area; R_2 = programming language; R_3 = commits; and R_4 = additions.

The next step consists of labeling the overall rating as the output of the ANN formula as shown in Figure 12.10, and after that get the intersection between the user request and the rated list of sub-repositories containing the relevant requirements document by making use of Equation (12.4). Considering the common words on both sides of users' requests and the sub-repository description gotten after applying NLP functions, we used Equation (12.5) to determine the satisfaction level of the user's expectation.

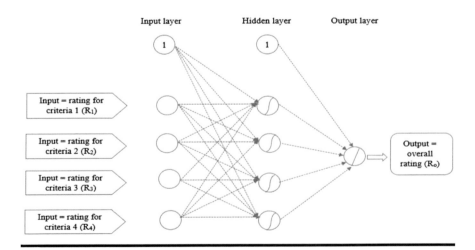

Figure 12.10   A conceptual view of ANN-based SAMLRS.

$$Sim = \frac{Ur \cap Prd}{Prd} \quad Sim = \frac{Ur \cap Prd}{Prd} \tag{12.4}$$

$$L(rd) = invoke(r) * \frac{cluster(ro)}{Max(clustering)} \quad L(rd) = invoke(r) * \frac{cluster(ro)}{Max(clustering)} \tag{12.5}$$

where *Ur* represents the user request, and *Prd* represents the project requirements document and *L(rd)* is the recommended list function. The cluster is a function that groups items under a category. The invoke(*r*) is the expert rating on a repository. Based on Equation (12.5), the algorithm in Figure 12.11 was used to achieve the Top-N recommendation.

12.6 System Implementation and Performance Evaluation

We implemented a prototype tool called System Analysis and Mining List Recommendation System (SAMLRS) using a Python web framework called Django to build the interfaces of the system, MySQL database and Python programming language. With the Python web-framework, the system can be made to run the phone, laptops, tablets and so on. Also, HTML and Bootstrap were used as front-end to build user views and Django as backend tools to perform the execution of the computational model. We ensured a user-friendly interface, as shown in Figures 12.12, 12.13 and 12.14, respectively, that can be used for recommendation purposes, which can be accessed from a browser in any operating system.

Going further, we evaluated the performance of the model using recall, precision and execution time as parameters. Concerning the recall and precision, we used an integrated python algorithm to determine the performance rate. The recall and precision are the proportion of positive cases. More precisely, precision determines the percentage of relevant results, while the recall determines the overall (total) results classified correctly. Thus, recall and precision help in determining the accuracy of our model. The number of false-positive (FP) and true positive (TP), and the number of true negative (TN) and the false negative (FN) were determined (see Table 12.3).

The performance evaluation was based on the 1,000 projects in the repository. Equations 12.6 and 12.7 were used for determining the recall and precision, respectively.

$$Recall = \frac{TP}{TP + FN} \quad Recall = \frac{642}{642 + 95} \quad Recall = 0.871099 \tag{12.6}$$

$$Precision = \frac{TP}{TP + FP} \quad Precision = \frac{624}{642 + 118} \quad Precision = 0.844736 \tag{12.7}$$

```
Algorithm: System_recommendation|

  Get["request"]
    tokens = word_tokenize("request")
    f == tokens
    stop_words = set(stopwords.words('english'))
    filtered_sentence = [w for w in f if not w in stop_words]
    filtered_sentence = []
    for w in f:
      if w not in f:
        filtered_sentence.append(w)
    l==Get["domain"]
    m==Get["pl"]

    import mysql.connector
    from mysql.connector import Error
 try:
    mySQLconnection = mysql.connector.connect(host='localhost',
                    database='msr',
                    user='root',
                    password=' ')
    sql_select_Query = "select pl, domain from dataset where (pl == m) && (domain== l)"
    cursor = mySQLconnection .cursor()
    cursor.execute(sql_select_Query)
    Y == sql_select_Query
    records = cursor.fetchall()
    w == k.length() <!--number of word in processed in request-->
    n == l.length() <!--number of word in processed in requproject description -->
 for: x = 1; x<= Y
  for: i = 1; i<= w; i++ do
    for: j = 1; j<= n; i++ do
        String s = ' '
        if: m < n then
           v == sim(i,j)
        end if
    end for
  end for
 s =v+;
 if: s>=0.8 then
 lp = project
 end if
 return  lp
 end for
```

Figure 12.11   Top-N recommendation algorithm.

Figure 12.12   Home page interface of SAMLRS.

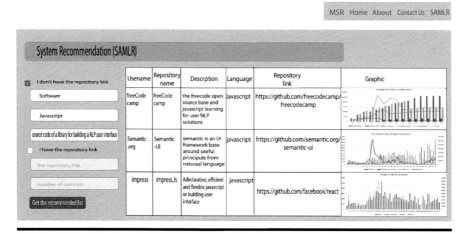

Figure 12.13   User page interface of SAMLRS.

Thus, 87% of recall and 84% of precision was accomplished with our developed model. We used the "*timeit*" library to determine the execution time of our approach. The algorithm in Figure 12.15, and the sequence of code display in Figure 12.16 were used to determine the execution time.

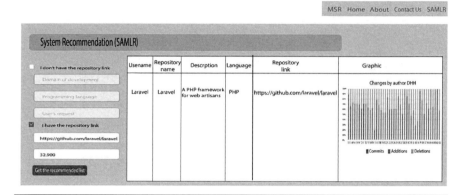

Figure 12.14   SAMLRS result using repository link.

Table 12.3   Metrics Table Showing the Analysis of TP, FP, FN, and TN in the Dataset

	Relevant	*Non-Relevant*
Retrieve	TP = 642	FP = 118
Non-Retrieve	FN = 95	TN = 145

```
Algorithm: Execution time
def timeit(method):
    def timed(*args, **kw)
        ts = time.time()
        result = method(*args, **kw)
        te = time.time()
        if 'log_time' in kw:
            name = kw.get('log_time, method.__name__.upper())
            kw['log_time'][name] = int((te − ts) * 1000)
        else:
            print '%r %2.2f ms' % \
                (method.__name__, (te − ts) * 1000)
    Return result
```

Figure 12.15   Algorithm for determining the execution time.

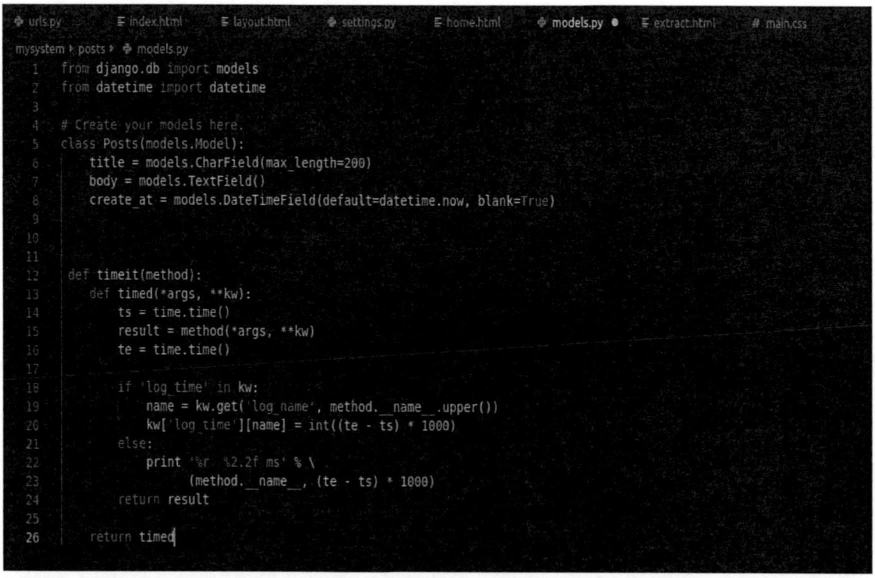

```
 urls.py     index.html     layout.html     settings.py     home.html     models.py ●     extract.html     # main.css
mysystem ▸ posts ▸  models.py
  1   from django.db import models
  2   from datetime import datetime
  3
  4   # Create your models here.
  5   class Posts(models.Model):
  6       title = models.CharField(max_length=200)
  7       body = models.TextField()
  8       create_at = models.DateTimeField(default=datetime.now, blank=True)
  9
 10
 11
 12   def timeit(method):
 13       def timed(*args, **kw):
 14           ts = time.time()
 15           result = method(*args, **kw)
 16           te = time.time()
 17
 18           if 'log_time' in kw:
 19               name = kw.get('log_name', method.__name__.upper())
 20               kw['log_time'][name] = int((te - ts) * 1000)
 21           else:
 22               print '%r  %2.2f ms' % \
 23                   (method.__name__, (te - ts) * 1000)
 24           return result
 25
 26       return timed
```

Figure 12.16 Sequence of code executed to determine the execution time.

12.7 Results and Discussion

With our approach, applicable and consistent output with a better view of the movement of data and their various structures could be achieved. Our research identified and mined the author's commits, additions, changes and deletions from the repositories. Data were labeled and categorized according to their importance. The graph in Figure 12.17 is an overview of the main components mined that carried most of the hidden and useful data in the repository. We discovered that over the years, the most active records in the repositories are essential for supporting software evolution. Also, the more the occurrence of commits on a software project development, the more successful the development project becomes.

The X and Y axes in Figure 12.17 represent the year from 2004 to 2018 and the number of "main components" changes, respectively. Changes per commits are like errors that were observed with time, and addition is the contribution made on errors. The more the commits in a project, the lesser the errors; in other words, the solution for a project is just with time when commits happen to make the project better. Moreover, there is some project produced by authors that did not get to the end or without a relevant solution. Figure 12.16 shows the data evolution in the repository with the quarterly Git commits changes. Figure 12.18 displays the total number of useful information to be recommended for locating the addition, changes and commits. As shown in Figure 12.18, from period 1 to 57 about 300,000 commits and changes are relevant to increase software project development. The author's log

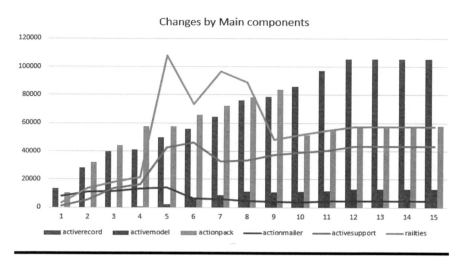

Figure 12.17 Changes by main components.

commits and contribution make a project more sustainable. The recommendation list was produced based on the users' requests. In Table 12.4, we show the breakdown of the author's log commits, that is, Git Commits Log-top Ten Contributors in 2019.

Figure 12.19 showed the structural flow of changes per week and what it is all about. Figures 12.20 and 12.21 display structured addition from an author and the flow of commits in ruby programming language, respectively. The final graph in Figure 12.22 shows the most frequently used programming languages during a software project development. The following table in the appendix section explains

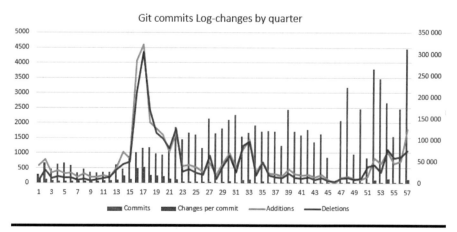

Figure 12.18 Git Commits Log-Changes by Quarter showing data evolution in the repository.

Table 12.4   Author Log Commits Showing Top Ten Contributors

	2019				
	Commits	*Additions*	*Deletions*	*Changes*	*Net Changes*
Rafael Mendonça França	776	2 145	3 212	5 357	-1 067
Aaron Patterson	432	4 375	4 093	8 468	282
Sean Griffin	414	6 772	4 339	11 111	2 433
Yves Senn	400	3 131	1 920	5 051	1 211
Kasper Timm Hansen	214	1 712	847	2 559	865
yui-knk	161	1 342	834	2 176	508
Zachary Scott	109	497	1 673	2 170	-1 176
yuuji.yaginuma	107	785	391	1 176	394
Xavier Noria	98	1 791	1 336	3 127	455
Ryuta Kamizono	97	2 820	1 938	4 758	882
All others	2 807	36 086	18 397	54 483	17 689

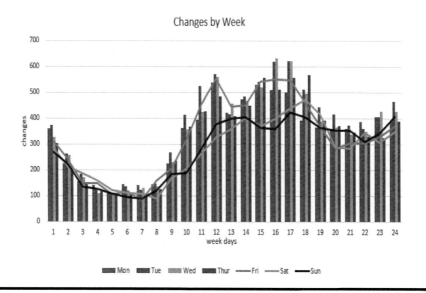

Figure 12.19   Changes by week and hours.

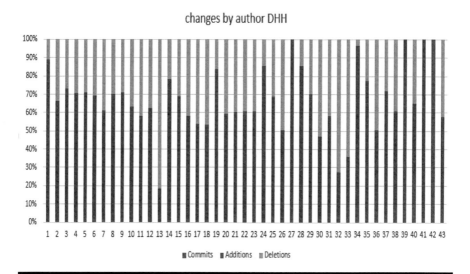

Figure 12.20   Changes by author DHH.

the data analysis: Table 12.5 shows changes that occurred by the "main components" from 2004 to 2018. Table 12.6 shows changes on an hourly and weekly
basis. Table 12.7 shows the most active pair that participated in sub-repositories
and the movement changes. An author (DHH) can add his contribution to the
repository as many times as possible, and intervene in a sub-repository to put his
change or commits (see Table 12.8 for the analysis of changes). Tables 12.9 and
12.10 show the Git commits Log-changes and the changes in Ruby file by quarter

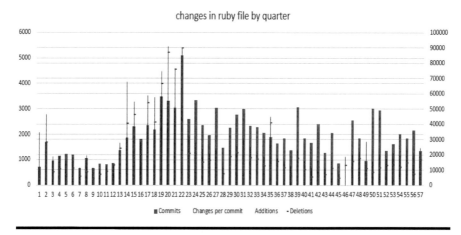

Figure 12.21   Changes in Ruby file on quarterly basis.

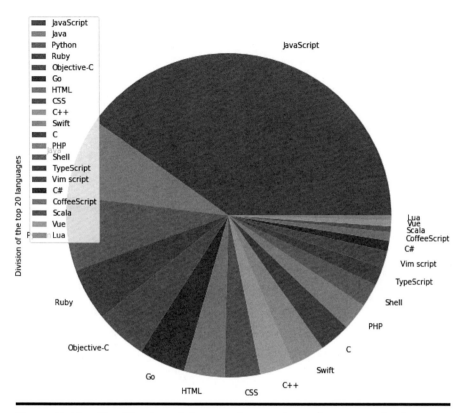

Figure 12.22 Classification of the most frequently used programming language in software project development.

respectively. In Table 12.11, we show the analysis of the most frequently used programming in software project development.

With the performance evaluation, the developed model was able to show a timely and up-to-date recommended list of requirements documents upon programmers' requests. With the row dataset, out of 1,000 repositories gotten from GitHub, more than 700 repositories were well structured. The system had 75% of performance in terms of structuring data in the repository and 84% of performance in terms of data recommendation with 1.98 seconds of execution time, as shown in Figure 12.23. The implication of our results means that programmers can locate functional requirements and design documents more effectively and efficiently.

Table 12.5 Changes by Main Components

Column 1	2004	2005	2006	2007	2008	2009	2010	2011	2012	2013	2014	2015	2016	2017	2018
activerecord	13247	28242	39827	40949	49631	55678	64609	76147	78488	85913	97172	105573	105573	105573	105573
activemodel	0	0	0	304	1998	6367	8680	11105	10911	11179	11619	13064	13064	13064	13064
actionpack	10331	31882	44037	57324	57558	65694	72338	78123	83814	51313	55300	58141	58141	58141	58141
actionmailer	8131	11003	11513	13352	14019	6489	5849	4481	4273	3788	4501	4746	4746	4746	4746
activesupport	1093	5340	13718	16372	42647	46155	32812	33350	37259	39074	40533	43319	43319	43319	43319
railties	3350	13748	17773	21237	107886	73391	96717	88795	48390	51880	54778	57496	57496	57496	57496

Table 12.6 Changes by Weekday and Hour

	00	01	02	03	04	05	06	07	08	09	10	11	12	13	14	15	16	17	18	19	20	21	22	23
Mon	361	226	154	142	110	143	142	143	226	363	396	539	422	473	530	510	501	394	366	361	360	388	407	391
Tue	375	263	186	127	111	135	123	146	268	415	526	572	417	485	542	620	624	511	445	416	374	361	406	465
Wed	328	257	172	108	112	120	130	136	232	358	425	560	458	469	520	633	622	497	391	366	339	350	428	428
Thur	303	214	144	116	109	106	105	126	237	368	428	486	408	451	559	513	558	570	394	371	347	332	326	390
Fri	313	244	148	148	104	116	83	159	204	319	450	553	447	451	542	550	548	460	369	283	305	306	330	369
Sat	287	213	186	158	120	109	108	90	166	205	262	328	359	402	370	399	439	472	417	290	283	345	308	348
Sun	271	224	134	126	108	95	88	120	182	189	281	378	400	406	364	360	425	404	364	353	354	310	342	408

Table 12.7   Importance of Changes/Most Active Pairs

	Commits	Changes	Net Changes
Yehuda Katz + Carl Lerche	183	81 808	2 542
José Valim and Mikel Lindsaar	59	9 362	920
Jason Noble & Ralph Shnelvar	20	1 048	490
Jose and Yehuda	17	2 694	1 586
Yehuda Katz and Carl Lerche	10	1 036	476
Santiago Pastorino and José Ignacio Costa	9	85	−7
Carl Lerche & Yehuda Katz	7	1 084	8
Amparo Luna + Guillermo Iguaran	5	313	49
Carlos Galdino + Rafael Mendonça França	5	305	171
Coraline Ada Ehmke + Aaron Patterson	4	126	10
Jonathan Dance + Gabriel Horner	4	287	99
Akira Matsuda + Koichi Sasada	3	13	11
Grant Hutchins & Peter Jaros	3	340	208
Rafael Mendonça França + Kassio Borges	3	24	−10
Steve Klabnik + Katrina Owen	3	192	88

Table 12.8   Changes by Author DHH

Quarter	Commits	Additions	Deletions
Qrt1 2008	289	41 208	5 056
Qrt2 2008	637	45 634	23 579
Qrt3 2008	357	17 479	6 557
Qrt4 2008	348	14 557	6 252
Qrt1 2009	165	5 504	2 294
Qrt2 2009	277	11 476	5 161
Qrt3 2009	101	4 165	2 699
Qrt4 2009	121	8 044	3 454
Qrt1 2010	75	1 927	806
Qrt2 2010	57	2 178	1 298
Qrt3 2010	83	5 190	3 786
Qrt4 2010	149	6 132	3 738
Qrt1 2011	95	2 098	9 765
Qrt2 2011	75	4 726	1 326
Qrt3 2011	75	2 066	967
Qrt4 2011	21	706	520
Qrt1 2012	103	2 894	2 546
Qrt2 2012	27	1 110	992
Qrt3 2012	3	205	40
Qrt1 2013	48	1 193	851
Qrt2 2013	49	646	452
Qrt3 2013	40	660	450
Qrt4 2013	16	72	57
Qrt1 2014	4	196	33
Qrt2 2014	12	217	103
Qrt3 2014	88	18 792	18 547

(Continued)

Table 12.8 (Continued)   Changes by Author DHH

Quarter	Commits	Additions	Deletions
Qrt4 2014	2	20	0
Qrt1 2015	17	563	97
Qrt2 2015	7	57	27
Qrt3 2015	20	595	694
Qrt4 2015	30	299	234
Qrt1 2016	18	94	297
Qrt2 2016	16	42	104
Qrt3 2016	3	26	1
Qrt4 2016	35	400	127
Qrt1 2017	12	95	105
Qrt2 2017	84	1 300	542
Qrt3 2017	45	319	236
Qrt4 2017	6	1	0
Qrt1 2018	18	25	23
Qrt2 2018	3	109	0
Qrt3 2018	3		
Qrt4 2018	1	50	37

Table 12.9 Git Commits Log-Changes by Quarter

Quarters	Total Lines	Commits	Changes Per Commit	Additions	Deletions
Qrt4 2004	36 152	289	160,08	41 208	5 056
Qrt1 2005	59 371	664	128,5	54 273	31 054
Qrt2 2005	72 421	470	73,14	23 714	10 664
Qrt3 2005	87 931	630	70,21	29 872	14 362
Qrt4 2005	98 221	664	53,27	22 830	12 540
Qrt1 2006	108 996	589	61,52	23 504	12 729
Qrt2 2006	115 612	338	59,66	13 391	6 775
Qrt3 2006	128 229	474	66,5	22 070	9 453
Qrt4 2006	136 824	367	49,13	13 313	4 718
Qrt1 2007	141 534	347	61,58	13 039	8 329
Qrt2 2007	148 045	372	76,78	17 537	11 026
Qrt3 2007	148 195	364	76,37	13 975	13 825
Qrt4 2007	153 035	609	109,3	35 703	30 863
Qrt1 2008	180 483	465	248,48	71 495	44 047
Qrt2 2008	188 308	890	119,03	56 882	49 057
Qrt3 2008	262 450	1 015	486,93	284 186	210 044
Qrt4 2008	279 228	1 160	539,32	321 193	304 415
Qrt1 2009	250 391	1 187	261,37	140 702	169 539
Qrt2 2009	260 540	974	248,23	125 962	115 813
Qrt3 2009	269 607	940	229,19	112 254	103 187
Qrt4 2009	261 142	1 086	141,26	72 470	80 935
Qrt1 2010	252 265	1 857	131,45	117 614	126 491
Qrt2 2010	264 814	1 450	46,07	39 675	27 126
Qrt3 2010	274 959	1 668	43,86	41 648	31 503
Qrt4 2010	288 552	1 601	37,24	36 610	23 017
Qrt1 2011	293 720	1 162	37,55	24 403	19 235
Qrt2 2011	275 980	2 135	51,12	45 702	63 442
Qrt3 2011	285 781	1 641	18,25	19 873	10 072

(Continued)

Table 12.9 (Continued)   Git Commits Log-Changes by Quarter

Quarters	Total Lines	Commits	Changes Per Commit	Additions	Deletions
Qrt4 2011	292 841	1 819	47,13	46 396	39 336
Qrt1 2012	298 964	2 107	62,82	69 245	63 122
Qrt2 2012	310 371	2 266	27,12	36 436	25 029
Qrt3 2012	302 343	1 554	105,25	77 763	85 791
Qrt4 2012	307 578	1 672	117,73	101 039	95 804
Qrt1 2013	314 854	1 925	21,72	24 541	17 265
Qrt2 2013	316 653	1 708	56,88	49 478	47 679
Qrt3 2013	320 199	1 735	22,73	21 489	17 943
Qrt4 2013	327 079	1 712	20,02	20 578	13 698
Qrt1 2014	331 814	1 247	22,35	16 301	11 566
Qrt2 2014	340 761	2 450	22,04	31 469	22 522
Qrt3 2014	348 884	1 708	19,39	20 620	12 497
Qrt4 2014	355 046	1 586	18,08	17 419	11 257
Qrt1 2015	361 582	1 764	18,59	19 667	13 131
Qrt2 2015	366 769	1 359	16,5	13 808	8 621
Qrt3 2015	374 022	1 631	19,46	19 495	12 242
Qrt4 2015	377 522	861	15,65	8 486	4 986
Qrt1 2016	384 058	1 250	17,25	9 845	3 258
Qrt2 2016	391 160	2 086	20,08	12 354	10 654
Qrt3 2016	396 322	3 184	34,58	15 650	12 058
Qrt4 2016	440 108	965	12,15	8 923	7 235
Qrt1 2017	455 792	2 456	20,56	7 654	10 237
Qrt2 2017	461 204	844	17,02	15 237	37 486
Qrt3 2017	487 746	3 785	98,95	58 487	41 546
Qrt4 2017	588 760	3 457	25,32	45 125	25 457
Qrt1 2018	599 018	2 658	152,12	72 378	78 248
Qrt2 2018	603 343	1 547	23,81	45 457	58 354
Qrt3 2018	605 401	2 457	30,54	50 568	61 345
Qrt4 2018	617 307	4 457	125,87	125 234	75 487

Table 12.10   Changes in Ruby File by Quarter

	Total Lines	Commits	Changes Per Commit	Additions	Deletions
Qrt4 2004	30 167	709	54,83	34 522	4 355
Qrt1 2005	48 571	1 700	43,85	46 476	28 072
Qrt2 2005	58 694	949	28,37	18 525	8 402
Qrt3 2005	66 847	1 161	24,77	18 455	10 302
Qrt4 2005	75 369	1 225	17,85	15 195	6 673
Qrt1 2006	82 995	1 214	23,97	18 361	10 735
Qrt2 2006	88 547	661	25,59	11 233	5 681
Qrt3 2006	99 379	1 063	26,39	19 441	8 609
Qrt4 2006	106 801	657	22,70	11 169	3 747
Qrt1 2007	110 198	831	20,46	10 198	6 801
Qrt2 2007	114 142	808	27,10	12 921	8 977
Qrt3 2007	111 542	864	27,03	10 377	12 977
Qrt4 2007	114 893	1 378	37,47	27 494	24 143
Qrt1 2008	141 902	1 878	57,50	67 500	40 491
Qrt2 2008	150 574	2 310	43,63	54 725	46 053
Qrt3 2008	159 467	1 811	26,62	28 550	19 657
Qrt4 2008	164 246	2 361	47,73	58 736	53 957
Qrt1 2009	180 349	2 185	45,11	57 329	41 226
Qrt2 2009	188 561	3 488	40,44	74 630	66 418
Qrt3 2009	193 048	3 323	53,53	91 190	86 703
Qrt4 2009	184 365	3 038	46,97	67 006	75 689
Qrt1 2010	166 093	5 092	31,59	71 292	89 564
Qrt2 2010	177 191	2 613	20,16	31 891	20 793
Qrt3 2010	179 910	3 333	16,01	28 033	25 314
Qrt4 2010	188 331	2 361	16,27	23 414	14 993
Qrt1 2011	193 023	1 962	18,71	20 705	16 013
Qrt2 2011	192 390	3 050	15,11	22 726	23 359
Qrt3 2011	197 937	1 483	13,27	12 616	7 069

(Continued)

Table 12.10 (Continued)   Changes in Ruby File by Quarter

	Total Lines	Commits	Changes Per Commit	Additions	Deletions
Qrt4 2011	202 738	2 265	18,43	23 275	18 474
Qrt1 2012	207 039	2 763	16,89	25 487	21 186
Qrt2 2012	216 879	2 988	16,86	30 107	20 267
Qrt3 2012	224 064	2 325	18,44	25 025	17 840
Qrt4 2012	228 216	2 286	16,79	21 266	17 114
Qrt1 2013	234 372	2 057	15,22	18 733	12 577
Qrt2 2013	238 517	1 899	45,16	44 951	40 806
Qrt3 2013	240 503	1 639	19,85	17 259	15 273
Qrt4 2013	244 285	1 848	14,38	15 180	11 398
Qrt1 2014	250 536	1 370	13,51	12 382	6 131
Qrt2 2014	256 843	3 064	13,34	23 592	17 285
Qrt3 2014	262 672	1 851	14,08	15 946	10 117
Qrt4 2014	269 044	1 680	9,75	11 380	5 008
Qrt1 2015	273 610	2 406	11,51	16 135	11 569
Qrt2 2015	277 707	1 267	14,11	10 988	6 891
Qrt3 2015	283 659	2 069	13,32	16 754	10 802
Qrt4 2015	286 516	849	13,86	7 312	4 455
Qrt1 2016	293 460	1 847	20,56	18 576	12932
Qrt2 2016	297 007	2 546	18,87	20 254	15 358
Qrt3 2016	303 385	1 847	25,12	21 876	17 420
Qrt4 2016	313 310	947	12,38	28 542	10 875
Qrt1 2017	318 914	3 025	10,32	15 256	8 345
Qrt2 2017	325 278	2 952	14,07	20 672	11 201
Qrt3 2017	335 824	1 352	12,05	17 526	12 952
Qrt4 2017	344 368	1 625	14,16	20 553	13 287
Qrt1 2018	347 856	2 021	10,25	19 587	11 846
Qrt2 2018	358 660	1 845	13,04	22 658	14 654
Qrt3 2018	366 460	2 154	14,26	20 756	7 309
Qrt4 2018	376 798	1 345	10,23	24 234	16 687

Table 12.11   Most Used Programming Language in Software Project Development

Programming Language	Number of Programming Language Utilization in a Project
Javascript	900
Java	425
C++	126
PHP	94
CoffeeScript	42
C	100
Go	167
R	20
Ruby	189
Python	350
HTML	158
CSS	147
TypeScript	71
Shell	87
Swift	113
C#	55
Vim Script	66
Scala	31
Vue	24
Lua	18
Objective-C	257

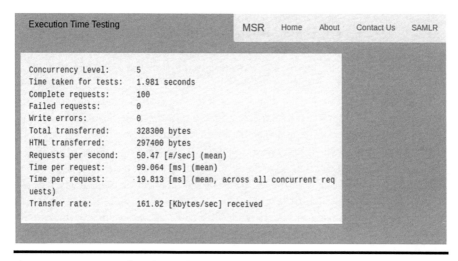

Figure 12.23   Execution time of a request demand.

12.8 Conclusion

In this chapter, we focused on the development of a model for mining software repositories to discover hidden requirements and design documents. The goal is to make a recommendation of the most matched with users' requests. We adopted a quantitative research approach by experimenting with a developed tool called SAMLRS. We extracted data from GitHub to evaluate software development practices in the GitHub repository. We focused on the GitHub repository that contains up to a million of sub-repository considered as open group project development where more than eleven million users upload and access information into the repositories. Noticeably, mining unstructured SR has excellent advantages for SE practices in terms of project development (Thomas, 2012). Nevertheless, solving the problem of complexity is tasking because of the unstructured nature of data, noises in the data and unlabeled data in the repositories.

We observed that the occurrence of these complexities does not go well with software developers because it makes them waste lots of time and effort to produce software products during project development. Thus, we identified unlabeled data as noises by using natural language text that contains special characters and English words. By noises, we mean data that has wrong outliers and null information not significant to the dataset. We used NLP to handle the noises in the dataset. Also, we grouped the text according to their content and labels before storing it in a database by applying the machine learning technique. Thus, our research provided a model that can be used during a software development project to reduce the complexity and noises of data in software repositories. Practically this work has contributed to knowledge by providing a system that recommends requirement documents related to source code to software developers according to their request.

In the future, we observed the need for further collaborations among software engineers to develop a system that will consider all the repositories that work on requirements and design documents. In practice, developers focused majorly on source code without paying attention to requirements specification and architectural design. Regrettably, no implementation can be done without having the precise requirement and design specifications. So, a system that will integrate other repositories, requirements and design documents are expected. The implication of this will mean that actors of another repository may be able to access proper information efficiently and timely, and then produce a quality system. In the long term, a solution should be found, for which software engineers and developers can have a mined repository that will be used for collecting and mining data as soon as changes are made in the repository. Also, hybridizing machine learning and NLP techniques have the potential of engendering better training on the dataset without necessary coding as the case is with existing techniques. Finally, our research can serve as a guide to other researchers for future development within "real-time mined repository" tasks.

References

Abebe, S. L., Ali, N., & Hassan, A. E. (2016). An empirical study of software release notes. *Empirical Software Engineering*, *21*(3), 1107–1142.

Agrawal, A., Fu, W., & Menzies, T. (2018). What is wrong with topic modelling? and how to fix it using search-based software engineering. *Information and Software Technology*, *98*, 74–88.

Arellano, A., Carney, E., & Austin, M. A. (2015, April). Natural language processing of textual requirements. In *The Tenth International Conference on Systems (ICONS 2015), Barcelona, Spain* (pp. 93–97).

Arnaoudova, V., Di Penta, M., Antoniol, G., & Gueheneuc, Y. G. (2013). A new family of software anti-patterns: Linguistic anti-patterns. In *17th IEEE European Conference on Software Maintenance and Reengineering, Genova, Italy, 5-8 March 2013*, (pp. 187–196).

Bavota, G. (2016). Mining unstructured data in software repositories: Current and future trends. In *23rd IEEE International Conference Proceedings on Software Analysis, Evolution, and Reengineering (SANER), Osaka, Japan, 14-18 March 2016*, (Vol. 5, pp. 1–12).

Bavota, G., Oliveto, R., Gethers, M., Poshyvanyk, D., & De Lucia, A. (2013). Method book: Recommending move method refactorings via relational topic models. *IEEE Transactions on Software Engineering*, *40*(7), 671–694.

Bird, S., Klein, E., & Loper, E. (2009). *Natural language processing with Python: Analyzing text with the natural language toolkit*. "O'Reilly Media, Inc."

Buse, R. P., & Weimer, W. (2010). Automatically documenting program changes. In *ASE* (Vol. 10, pp. 33–42).

Buse, R. P., & Zimmermann, T. (2012). Information needs for software development analytics. In Proceedings of the IEEE 34th International Conference on Software Engineering (ICSE) (pp. 987–996).

Canfora, G., & Cerulo, L. (2005, September). Impact analysis by mining software and change request repositories. In *Proceedings of the IEEE 11th International Software Metrics Symposium (METRICS'05)* (pp. 20–29).

Chan, W. K., Cheng, H., & Lo, D. (2012). Searching connected API subgraph via text phrases. *In* Proceedings of the ACM SIGSOFT 20th International Symposium on the Foundations of Software Engineering (p. 10).

Chaturvedi, K. K., & Singh, V. B. (2012, September). Determining bug severity using machine learning techniques. In *2012 CSI Sixth International Conference on Software Engineering (CONSEG)* (pp. 1–6). IEEE.

Chen, T. H., Thomas, S. W., & Hassan, A. E. (2016). A survey on the use of topic models when mining software repositories. *Empirical Software Engineering, 21*(5), 18431919.

Cordeiro, J., Antunes, B., & Gomes, P. (2012). Context-based recommendation to support problem-solving in software development. In *2012 Third International Workshop on Recommendation Systems for Software Engineering (RSSE)* (pp. 85–89). IEEE.

Cortés-Coy, L. F., Linares-Vásquez, M., Aponte, J., & Poshyvanyk, D. (2014). On automatically generating commit messages via summarization of source code changes. In *14th IEEE International Working Conference Proceedings on Source Code Analysis and Manipulation, Victoria, British Columbia, Canada, 28–29 September 2014,* (pp. 275–284).

Elsen, E., Hill, E., Pollock, L., & Vijay-Shanker, K. (2009). Mining source code to automatically split identifiers for software analysis. *IEEE Xplore.*

Goldman, M., & Miller, R. C. (2009). Codetrail: Connecting source code and web resources. *Journal of Visual Languages & Computing, 20*(4), 223–235.

Gousios, G., Pinzger, M., & Deursen, A. van. (2014). An exploratory study of the pull-based software development model. In *36th International Conference Proceedings on Software Engineering (ICSE'14), Hyderabad, India, May 31 to June 7, 2014,* pp 345–355.

Güemes-Peña, D., López-Nozal, C., Marticorena-Sánchez, R., & Maudes-Raedo, J. (2018). Emerging topics in mining software repositories. *Progress in Artificial Intelligence, 7*(3), 237–247.

Gupta, S., & Gupta, S. K. (2019a). Natural language processing in mining unstructured data from software repositories: A review. *Sādhanā, 44*(12), 244.

Gupta, S., & Gupta, S. K. (2019b). Abstractive summarization: An overview of state of the art. *Expert Systems with Applications, 121,* 49–65.

Haiduc, S., Kobayashi, T., Lanza, M., & Marcus, A. (2016). Mining & Modeling Unstructured Data in Software-Challenges for the Future.

Hassan, A. E. (2006). Mining software repositories to assist developers and support managers. *In Software Maintenance, 2006. ICSM'06. 22nd IEEE International Conference on* (pp. 339–342). *IEEE.*

Hassan, A. E. (2008). The road ahead for mining software repositories. In *2008 Frontiers of Software Maintenance* (pp. 48–57). IEEE.

Hassan, A. E., & Xie, T. (2010). Software intelligence: The future of mining software engineering data. In *Proceedings of the FSE/SDP Workshop on Future of Software Engineering Research* (pp. 161–166). ACM.

Hattori, L., dos Santos Jr, G., Cardoso, F., & Sampaio, M. (2008, October). Mining software repositories for software change impact analysis: A case study. In Proceedings of the 23rd Brazilian Symposium on Databases (pp. 210–223).

Junior, L. D. D., & Favero, E. (2012). Integrating software repository mining: A decision support centred approach. *International Journal of Software Engineering & Applications, 3*(6), 57.

Kaiya, H., & Saeki, M. (2006, September). Using domain ontology as domain knowledge for requirements elicitation. In *14th IEEE International Requirements Engineering Conference (RE'06)* (pp. 189–198). IEEE.

Kalliamvakou, E., Gousios, G., Blincoe, K., Singer, L., German, D. M., & Damian, D. (2014). The promises and perils of mining GitHub. In Proceedings of the 11th ACM Working Conference on Mining Software Repositories (pp. 92–101).

Kasianenko, S. (2018). Predicting Software Defectiveness by Mining Software Repositories. *Masters Thesis, Linnaeus University, Faculty of Technology, Department of Computer Science and Media Technology (CM).*

Kemerer, C. F., & Slaughter, S. (1999). An empirical approach to studying software evolution. *IEEE Transactions on Software Engineering, 25*(4), 493–509.

Kharlamov, A. A., Orekhov, A. V., Bodrunova, S. S., & Lyudkevich, N. S. (2019, December). Social network sentiment analysis and message clustering. In *International Conference on Internet Science,* Springer, Cham, (pp. 18–31).

Kiefer, C., Bernstein, A., & Tappolet, J. (2007). Mining software repositories with ISPAROL and a software evolution ontology. *In IEEE Computer Society, Fourth International Workshop Proceedings on Mining Software Repositories*(p. 10).

Lindvall, M., Rus, I., Jammalamadaka, R., & Thakker, R. (2001*). Software tools for knowledge management: A DACS State-of-the-art report. Fraunhofer Center for Experimental Software Engineering, Maryland, USA.*

Lotufo, R., Malik, Z., & Czarnecki, K. (2015). Modelling the 'hurried' bug report reading process to summarize bug reports. *Empirical Software Engineering, 20*(2), 516–548.

Mani, S., Catherine, R., Sinha, V. S., & Dubey, A. (2012). Ausum: Approach for unsupervised bug report summarization. In *Proceedings of the ACM SIGSOFT 20th International Symposium on the Foundations of Software Engineering* (p. 11). ACM.

Manning, C., Surdeanu, M., Bauer, J., Finkel, J., Bethard, S., & McClosky, D. (2014, June). The Stanford CoreNLP natural language processing toolkit. In *Proceedings of 52nd Annual Meeting of the Association for Computational Linguistics: System Demonstrations* (pp. 55–60).

Marcus, A., Poshyvanyk, D., & Ferenc, R. (2008). Using the conceptual cohesion of classes for fault prediction in object-oriented systems. *IEEE Transactions on Software Engineering, 34*(2), 287–300.

Menzies, T., Greenwald, J., & Frank, A. (2006). Data mining static code attributes to learn defect predictors. *IEEE Transactions on Software Engineering, 33*(1), 2–13.

Moreno, L., Bavota, G., Di Penta, M., Oliveto, R., & Marcus, A. (2015, May). How can I use this method? In *2015 IEEE/ACM 37th IEEE International Conference on Software Engineering* (Vol. 1, pp. 880–890). IEEE.

Moreno, L., Bavota, G., Di Penta, M., Oliveto, R., Marcus, A., & Canfora, G. (2014). Automatic generation of release notes. In *Proceedings of the 22nd ACM SIGSOFT International Symposium on Foundations of Software Engineering* (pp. 484–495). ACM.

Nazir, F., Butt, W. H., Anwar, M. W., & Khattak, M. A. K. (2017). The applications of natural language processing (NLP) for software requirement engineering—a systematic literature review. In *International Conference on Information Science and Applications* (pp. 485–493). Springer, Singapore.

Panichella, A., Dit, B., Oliveto, R., Di Penta, M., Poshynanyk, D., & De Lucia, A. (2013, May). How to effectively use topic models for software engineering tasks? An approach based on genetic algorithms. In *2013 35th International Conference on Software Engineering (ICSE)* (pp. 522–531). IEEE.

Polyakov, E. V., Voskov, L. S., Abramov, P. S., & Polyakov, S. V. (2020). Generalized approach to sentiment analysis of short text messages in natural language processing. *ИНфОРМАциОННО-УпРАВляющиЕ СиСТЕМы,* (1), 2–14.

Poncin, W., Serebrenik, A., & Van Den Brand, M. (2011). Process mining software repositories. *In 15th European Conference on Software Maintenance and Reengineering (CSMR), 2011* (pp. 5–14). IEEE.

Ponzanelli, L., Bavota, G., Di Penta, M., Oliveto, R., & Lanza, M. (2014). Mining stack overflow to turn the IDE into a self-confident programming prompter. In *Proceedings of the 11th Working Conference on Mining Software Repositories* (pp. 102–111). ACM.

Poshyvanyk, D., & Marcus, A. (2006). The conceptual coupling metrics for object-oriented systems. In *2006 22nd IEEE International Conference on Software Maintenance* (pp. 469–478). IEEE.

Prakash, B. A., Ashoka, D. V., & Aradhya, V. M. (2012). Application of data mining techniques for software reuse process. *Procedia Technology, 4,* 384–389.

Rastkar, S., & Murphy, G. C. (2013). Why did this code change? In *Proceedings of the 2013 International Conference on Software Engineering* (pp. 1193–1196). IEEE Press.

Rastkar, S., Murphy, G. C., & Murray, G. (2010). Summarizing software artefacts: A case study of bug reports. In *2010 ACM/IEEE 32nd International Conference on Software Engineering* (Vol. 1, pp. 505–514). IEEE.

Rigby, P. C., & Robillard, M. P. (2013). Discovering essential code elements in informal documentation. In *2013 35th International Conference on Software Engineering (ICSE)* (pp. 832–841). IEEE.

Rodriguez, D., Herraiz, I., & Harrison, R. (2012, June). On software engineering repositories and their open problems. In *2012 First International Workshop on Realizing AI Synergies in Software Engineering (RAISE)*(pp. 52–56). IEEE.

Ruparelia, N. B. (2010). Software development lifecycle models. *ACM SIGSOFT Software Engineering Notes, 35*(3), 8–13.

Shi, W., Sun, X., Li, B., Duan, Y., & Liu, X. (2015). Using feature-interface graph for automatic interface recommendation: A case study. In IEEE Third International Conference on Advanced Cloud and Big Data, 2015 (pp. 296–303).

Siddiqui, T., & Ahmad, A. (2018). Data mining tools and techniques for mining software repositories: A systematic review. In *Big Data Analytics* (pp. 717–726). Springer, Singapore.

Steinmacher, I., Conte, T., Gerosa, M. A., & Redmiles, D. (2015). Social barriers faced by newcomers placing their first contribution in open-source software projects. In Proceedings of the 18th ACM Conference on Computer Supported Cooperative Work & Social Computing (pp. 1379–1392).

Storks, S., Gao, Q., & Chai, J. Y. (2019). Commonsense reasoning for natural language understanding: A survey of benchmarks, resources, and approaches. *arXiv preprint arXiv:1904.01172.*

Stylos, J., & Myers, B. A. (2006). Mica: A web-search tool for finding API components and examples. In *Visual Languages and Human-Centric Computing (VL/HCC'06)* (pp. 195–202). IEEE.

Subramanian, S., Inozemtseva, L., & Holmes, R. (2014). Live API documentation. In *Proceedings of the 36th International Conference on Software Engineering* (pp. 643–652). ACM.

Sun, X., Li, B., Duan, Y., Shi, W., & Liu, X. (2016*). Mining software repositories for automatic interface recommendation.* Scientific Programming.

Takhteyev, Y., & Hilts, A. (2010). Investigating the geography of open source software through GitHub. *Manuscript submitted for publication.*

Takuya, W., & Masuhara, H. (2011). A spontaneous code recommendation tool based on associative search. In *Proceedings of the 3rd International Workshop on Search-Driven Development: Users, Infrastructure, Tools, and Evaluation* (pp. 17–20). ACM.

Thomas, S. (2012). Mining unstructured software repositories using IR models (Doctoral dissertation) from the School of Computing at Queen's University.

Thomas, S. W. (2011). Mining software repositories using topic models. In *Proceedings of the 33rd International Conference on Software Engineering* (pp. 1138–1139). ACM.

Thomas, S. W., Hassan, A. E., & Blostein, D. (2014). Mining unstructured software repositories. In *Evolving Software Systems* (pp. 139–162). Springer, Berlin, Heidelberg.

Thung, F., Wang, S., Lo, D., & Lawall, J. (2013). Automatic recommendation of API methods from feature requests. *In Proceedings of the 28th IEEE/ACM International Conference on Automated Software Engineering* (pp. 290–300). IEEE Press.

Valetto, G., Helander, M., Ehrlich, K., Chulani, S., Wegman, M., & Williams, C. (2007, May). Using software repositories to investigate socio-technical congruence in development projects. In *Fourth International Workshop on Mining Software Repositories (MSR'07: ICSE Workshops 2007)* (pp. 25–25). IEEE.

Wang, J., Dang, Y., Zhang, H., Chen, K., Xie, T., & Zhang, D. (2013). Mining succinct and high-coverage API usage patterns from source code. In Proceedings of the 10th IEEE Working Conference on Mining Software Repositories (pp. 319–328).

Wang, X., Zhang, L., Xie, T., Anvik, J., & Sun, J. (2008). An approach to detecting duplicate bug reports using natural language and execution information. In *Proceedings of the 30th international conference on Software engineering* (pp. 461–470). ACM.

Weiss, S. M., Indurkhya, N., Zhang, T., & Damerau, F. J. (2005). From textual information to numerical vectors. In *Text Mining* (pp. 15–46). Springer, New York, NY.

Xie, T., & Pei, J. (2006). MAPO: Mining API usages from open source repositories. In Proceedings of the ACM 2006 International Workshop on Mining Software Repositories (pp. 54–57).

Yousef, A. H. (2015). Extracting software static defect models using data mining. *Ain Shams Engineering Journal, 6*(1), 133–144.

Zhong, B., Li, H., Luo, H., Zhou, J., Fang, W., & Xing, X. (2020). Ontology-based semantic modeling of knowledge in construction: Classification and identification of hazards implied in images. *Journal of Construction Engineering and Management, 146*(4), 04020013.

Zhong, H., Xie, T., Zhang, L., Pei, J., & Mei, H. (2009). MAPO: Mining and recommending API usage patterns. In *European Conference on Object-Oriented Programming* (pp. 318–343). Springer, Berlin, Heidelberg.

Chapter 13

Empirical Studies on Using Pair Programming as a Pedagogical Tool in Higher Education Courses: A Systematic Literature Review

Kuljit Kaur Chahal[a], Amanpreet Kaur[a], and Munish Saini[a]

[a]*Guru Nanak Dev University, Amritsar, India*

Contents

13.1 Introduction 252
13.2 Related Work 253
13.3 Methodology...254
    13.3.1 Defining the Research Questions 255
    13.3.2 Choosing a Search Strategy and Devising Article Selection
        Criteria 255
        13.3.2.1 Data Sources 255
        13.3.2.2 Search Strategy and Selection Criteria.....................255
        13.3.2.3 Data Extraction...257
        13.3.2.4 Analysis and Classification 257

13.4 Results and Analysis 258
    13.4.1 What are the Measures of Interest to Analyze Effectiveness of
          Pair Programming in Computer Science Education? [RQ-1] 262
    13.4.2 What is the Effect of Pair Programming on These Measures
          of Interest? [RQ-2] 264
    13.4.3 What are Students' Perceptions of Using Pair Programming?
          [RQ-3] 266
    13.4.4 What are the Factors that Determine Effectiveness of PP?
          [RQ-4] 269
    13.4.5 Is Pair Programming Methodology Beneficial for Students'
          Teaming with Geographically Dispersed Partners? [RQ-5] 273
13.5 Threats to Validity 275
13.6 Issues and Future Work 275
13.7 Conclusion 278
Note 278
References 279

13.1 Introduction

Pair Programming (PP) is a practice that involves two programmers working together to accomplish a common programming task, literally sharing a keyboard and a computer (Williams et al., 2000). Typically, when one is writing code (a.k.a. the driver), the other one is observing and reviewing the work for possible defects and solutions (a.k.a. the navigator or observer). They should swap roles frequently during the programming sessions. Nosek (1998) did the first study to understand efficacy of the practice for professional programmers, and reported statistically significant better productivity, and quality of work of the experimental (pairs) group. Williams et al. (2000) executed the first experiment to test efficacy of PP in an education setting for the full duration of a semester in the University of Utah in 1999. The study reported statistically significant better quality of work of experimental (pairs) group, though their productivity was same as the control (solo) group. Looking at these benefits, people started experimenting with the practice in both the professional and academic environments.

PP has a variety of benefits for software professionals including better code quality, high productivity, and good attitude toward the practice (Hannay et al., 2009; da Silva Estacio & Prikladnicki, 2015). Several research studies indicate that PP is valuable in educational contexts as well (Hanks et al., 2011). The claimed benefits of PP in educational context include knowledge sharing, communication, greater enjoyment, less drop outs, and better performance in exams and programming assignments.

In order to apply the PP approach in classroom as educators or to investigate new interventions as researchers, we need to examine the existing experiences and state-of-the-art. However, the current state of the art of PP usage in education is under limited investigation. Although different aspects of PP in the educational context have been studied, we have not found a study in the recent times that examines in a systematic way the empirical work contributed in this domain. This article addresses the gap, and presents a systematized account of the empirical studies published on the PP practice in Computer Science education in universities/colleges. To accomplish this goal, this chapter presents results of a Systematic Literature Review (SLR) on the empirical studies of PP practice in higher education in the subject of Computer Science to systematically identify, analyze, and describe the state-of-art advances in the field.

In detail, the specific goals of this chapter are as follows:

■ To identify the measures of interest used in the empirical PP studies
■ To understand the effect of the PP practice on these measures of interest
■ To understand students' attitude toward the PP practice
■ To identify the factors that may impact the effectiveness of PP
■ To understand implications of the PP practice for students' teaming with geographically dispersed partners
■ To identify the research challenges and open questions to further improve the experimentation in the education context

The idea is not only to consolidate existing research but also to accelerate research in this direction as it is important to make computing enjoyable and improve retention rate in Computer Science due to universality of the desired computing skills in the 21st century. We review the literature published from 2000 to 2019 in major electronic databases. In order to assist Computer Science educators and support future research studies on this topic, this chapter reviews the current state of research on use of PP as a pedagogy tool. Specifically, we look at the different measures of interest to evaluate effectiveness of the PP practice along with the factors that may affect its effectiveness.

The chapter is organized as follows. Section 13.2 describes the related work. Section 13.3 presents the methods used to gather and select relevant articles, and describes how the review has been performed. Section 13.4 presents the results of the study, and answers the research questions. Section 13.5 discusses the threats to validity of the results and their mitigation strategies applied in the research methodology. Section 13.6 discusses issues in the existing work, and derives the future work opportunities. Section 13.7 concludes the chapter.

13.2 Related Work

The most recent work in this regard is a meta-analysis published in 2017 (Umapathy & Ritzhaupt, 2017). It analyzed the literature published from 2000 to 2014 (18 articles) and included only quantitative studies focusing on a limited set of measures related to students' performance and retention in programming courses. They conclude that PP is effective but it should be practiced with caution. Educators should follow the PP guidelines (Williams et al, 2008). We observed mixed results regarding the different measures of interest for evaluating PP effectiveness.

Prior to this, Salleh et al. (2010) reviewed the literature published from 1999 to 2007, and included studies focused on PP effectiveness on four broad categories of academic performance, technical productivity, quality, and satisfaction. They concluded that *time spent* was the most common measure of interest to test effectiveness of PP, and *skill level* was the most significant factor for making PP effective. However, this research found that students' performance using metrics like exam score is the most common measure of interest, and pair compatibility is the most significant factor determining PP effectiveness.

Hanks et al. (2011) reported a detailed review, not systematic, of the 10 years of PP usage in education field. Due to the mixed results of the PP experiments, Hanks et al. note lack of multi-institutional, longitudinal, and replication experiments of PP in education. This study also revealed the same. Unfortunately, the situation has not improved in the 10 years post their study. Researchers experimenting with the PP practice should give minute details so that their experiments can be replicated. Such detail is still missing in the literature. In this corpus of 68 articles, only one article (Rong et al., 2018) reports a replicated experiment. There are some other big concerns such as small sample size and lack of internal validity of the metrics used for evaluating efficacy of PP in education. Chen & Roy (2018) has addressed some of these concerns.

In the original PP approach, the programmer pair was considered to be collocated. Later on the PP methodology incorporated geographically dispersed pairs using their own devices connected through the Internet rather than sharing the same device and physical space. With considerable penetration of the Internet technology in our daily lives, it is important to look at the relevance of the PP technique for students who want to work on their programming assignments outside classroom. In the past, several empirical investigations have been conducted on effectiveness of PP in a distributed setup referred to as distributed pair programming (DPP), virtual pair programming (VPP), or remote pair programming (RPP). da Silva Estacio & Prikladnicki (2015) did a SLR on studies focused on the use of DPP in industry. This article aims at reporting a SLR of the empirical studies conducted on PP in the context of higher education including collocated as well as remote students.

13.3 Methodology

The study follows a review protocol (Kitchenham & Charters, 2007) to carry out the review in a systematic and structured manner. The review process included the following steps:

1. Defining the research questions
2. Choosing a search strategy and devising article selection criteria
3. Data extraction
4. Analysis and classification

The remainder of this section describes the steps by mentioning the procedures, the criteria, and application of the methodology.

13.3.1 Defining the Research Questions

The definition of research questions comprised the process of breaking down the main question into specific ones. These questions address strategic aspects of PP, and how studies explored these research questions.

RQ-1: What are the measures of interest in pair programming studies in the educational context?

RQ-2: How does pair programming affect these measures of interest?

RQ-3: What are students' perceptions of using pair programming?

RQ-4: What factors determine the effectiveness of pair programming?

RQ-5: Is pair programming methodology beneficial for students' teaming with geographically dispersed partners?

13.3.2 Choosing a Search Strategy and Devising Article Selection Criteria

This step consists of two phases: (1) defining and using a search strategy for identifying papers, and (2) deciding selection criteria for including/excluding papers.

13.3.2.1 Data Sources

The data sources in this study include electronic databases which index peer-reviewed conference/symposium proceedings, and academic journals. Following databases were explored for retrieving the papers: *IEEE Explore, ACM digital library, Springer Link, Science Direct, and Wiley online library.*

13.3.2.2 Search Strategy and Selection Criteria

To begin with, the authors developed a list of search terms to evaluate titles and abstracts. Database search started with a basic string agreed upon by all the authors. The string, given below, was used in the search process by applying the search string to the title, abstract, and keywords components of the articles in the electronic databases through the search window.

("pair programming") AND (teach OR learn OR education OR student OR course OR classroom OR virtual OR remote OR distributed OR online)

The search string includes terms to specify the concept ("pair programming"), and the scope of search (teach OR learn OR education OR student OR course OR classroom OR virtual OR remote OR distributed OR online). The synonyms were selected after consulting the literature in pilot searches.

Table 13.1 mentions the inclusion and exclusion criteria. The study includes articles that approach the use of classical PP (collocated as well as distributed) in higher education i.e. used by college/university students as the study subjects. The inclusion criteria (IC2) comprised of quality premises such as objectives, relevance, and rigor. Objectives narrowed down the scope of the study, and focused only relevant literature. Rigor filtered out articles published as abstracts only, posters, panel discussions, and short papers. Articles reporting findings in English are only included.

The exclusion criteria removed articles that study PP in K-12 education. It excluded articles of same authors with the same objective (EC2) and appearing in a conference as "initial results", and then appearing in a journal at a later date. In such cases, the authors retained the article published in a journal. EC3 excluded articles whose results were validated in replicated study. A replicated study provided insights into the earlier results as well. Literature reviews, or

Table 13.1   Inclusion and Exclusion Criteria

Reference	Criteria
Inclusion Criteria	
IC1	Articles that approach the use of classical pair programming (collocated as well as distributed) in education
IC2	Articles published as full paper in a conference, symposium, edited-book, or journal
IC3	Articles written in English
Exclusion Criteria	
EC1	Articles that used K-12 students as study subjects
EC2	Articles from same authors reporting same experiment with same objectives
EC3	Articles whose experiments were later replicated
EC4	Articles reporting literature reviews or systematic mapping studies
EC5	Full paper not available in electronic form
EC6	Articles using PP models different from the classical definition

systematic mapping studies were also dropped (EC4). EC5 dropped related articles published in offline mode as full papers are not available in electronic format, for example, articles published in Journal of Computing Sciences in Colleges. Finally, EC6 excludes articles not following the PP practice as per the definition, for example, partners using different computers (Gharoshi & Jensen, 2017).

Notably, there is no filter to select articles on the basis of date of publication. The publication year of the articles in this SLR ranges from 2000 to 2019. The review also includes papers irrespective of their analysis method unlike (Umapathy & Ritzhaupt, 2017), which includes only quantitative studies. Besides searching the electronic databases, the snowballing approach of finding any relevant research papers to augment the set was also applied by following the references given in the selected papers, and also searching the papers citing these primary set of papers. This multipronged approach was basically to reconnoiter all possible methods for finding relevant papers.

13.3.2.3 Data Extraction

The study used thematic analysis for understanding the underlying themes in the research literature. Thematic analysis is a method for identifying, analyzing, and reporting patterns (themes) within data. For the thematic analysis, this study used a bottom-up approach by creating codes from the paper content to identify and summarize the main points. Each author read 40% of the articles and recorded codes, and the overlapping subset of around 20% articles helped in establishing the reliability of the coding process. Any overlapping codes were combined after all agreed. The Cohen's kappa metric value 0.76 indicates the reliability of the process. Then the codes similar in nature were grouped and sub-themes were developed by summarizing the codes within each grouping that encapsulated the related codes. Thereafter, codes mentioning similar items were grouped and merged under an overarching theme. For example, "exam scores" and "programming assignment scores" were put together under the sub-theme of "student performance". Similarly, "class attendance" and "continuing with the subject in next course" were put under sub-theme "student retention". Then both the sub-themes merged under the overarching theme "Student Achievements" (see Table 13.2). Thematic analysis resulted in two overarching themes: Student achievements and Student perceptions.

13.3.2.4 Analysis and Classification

Thematic analysis resulted in the classification of articles into five groups aligned with the research questions. These groupings helped to organize the articles, with a

Table 13.2 Thematic Analysis for the Overarching Theme "Student Achievements"

Code	Sub-Theme	Main-Theme
Exam score	Performance	Student achievements
Programming assignment score		
Program correctness	Quality of work	
Defect density		
Code documentation		
Attendance	Retention	
Completed the course		
Continued with an advanced course		

focus on answering research questions. A thorough analysis of the selected papers helps to identify the following objectives that these articles stated to achieve:

- To investigate effectiveness of PP and the factors that may affect
- Does PP benefit students? Code quality, productivity, and lab attendance, and several other measures are used to compare experimental v/s control groups.
- To understand the students' perceptions of PP experiences
- How do students feel about the practice? Do they feel good or bad?
- To find the factors that may impact PP effectiveness
- These factors may include personality type, learning style, pair compatibility, and problem complexity.
- To investigate the PP practice for geographically dispersed students

13.4 Results and Analysis

This section describes the methodology execution, mentioning the number of papers selected for further analysis. It also discusses the results by finding answers to the five research questions. Table 13.3 illustrates the stage wise article selection workflow. At stage 1, the initial search retrieves the articles, by matching their title/abstract/keywords with the search string, from the electronic databases. As no filter for publication dates was applied during the search process, so the retrieved papers include all published at any date. Initially in the John Wiley and Sons electronic database, there were zero papers in the search results with the basic search string. While retrying by expanding the search by using only "pair programming"

Table 13.3 Number of Articles Selected at Each Stage

	Stage 1 Retrieved Using the Search String	Stage 2 Selected after Reading the Title and Abstract	Stage 3 Selected after Removing Duplicates Across Databases	Stage 4 Selected Following the Inclusion/ Exclusion Criteria
IEEE Explore	1363	84	84	23
ACM digital library	77	67	65	24
Springer	60	19	18	10
Science Direct	16	14	14	5
John Wiley and Sons	228	4	4	4
Others	-	-	-	2
Total	1744	188	185	68

as the search string in paper title, abstract, and keywords, the search results gave 228 papers. Stage 1 retrieved a total of 1744 articles, as of January 2020, from five different databases published during the period from 2000 through 2019. At stage 2 (see Table 13.3), the author responsible for searching a particular database shortlisted studies after reading title and abstract, and selected papers related to PP in educational context only. Accuracy of this step was cross verified by randomly checking 20% of the articles from each database.

Stage 3 removes duplicates across databases as some papers were indexed in multiple databases, for example, some papers were available in IEEE XPLORE as well as in ACM digital library. In case of jointly organized conferences by IEEE and ACM, papers were assigned to IEEE. Table 13.3 shows details of the numbers available in different databases at different stages of the search process. Out of the six papers found after using the sandboxing approach, only two were retained.

At the end of the search and filter process, following the inclusion/exclusion criteria in stage 4 (see Table 13.3), 68 articles relevant to the study made up the final list. Of these 68 articles, 52 are from conferences/Symposium/Book chapters and 16 are from journals. Table 13.4 shows a listing of the selected articles along with their unique identifiers. Each article is assigned a unique identifier to refer to the results in tabular format while discussing research questions in this section.

Table 13.4 List of Articles References in Chronological Order

Journal papers

ID	Reference	ID	Reference	ID	Reference	ID	Reference
[JP01]	Williams et al. (2000)	[JP05]	Hanks (2008)	[JP09]	Braught et al. (2011)	[JP13]	Erdei et al. (2017)
[JP02]	McDowell et al. (2006)	[JP06]	Simon & Hanks (2008)	[JP10]	Zacharis (2011)	[JP14]	Rong et al. (2018)
[JP03]	Müller (2007)	[JP07]	Choi et al. (2009)	[JP11]	Salleh et al. (2014)	[JP15]	Tsompanoudi et al. (2019)
[JP04]	Bipp et al. (2008)	[JP08]	Sfetsos et al. (2009)	[JP12]	Tsompanoudi et al. (2016)	[JP16]	Kavitha & Ahmed (2015)

Conference/Symposium Papers and Book Chapters

ID	Reference	ID	Reference	ID	Reference	ID	Reference
[CP17]	Gehringer (2003)	[CP30]	Slaten et al. (2005)	[CP43]	Radermacher & Walia (2011)	[CP56]	Nawahdah & Taji (2016)
[CP18]	Heiberg et al. (2003)	[CP31]	Hanks (2006)	[CP44]	Radermacher et al. (2012)	[CP57]	Benadé & Liebenberg (2017)
[CP19]	Janes et al. (2003)	[CP32]	Layman (2006)	[CP45]	Jones & Fleming (2013)	[CP58]	Rodríguez et al. (2017)
[CP20]	McDowell et al. (2003)	[CP33]	Melis et al. (2006)	[CP46]	Li et al. (2013)	[CP59]	Xinogalos et al. (2017)

(Continued)

Table 13.4 (Continued) List of Articles References in Chronological Order

Conference/Symposium Papers and Book Chapters								
ID	*Reference*	*ID*	*Reference*	*ID*	*Reference*	*ID*	*Reference*	
[CP21]	Nagappan et al. (2003)	[CP34]	Mendes et al. (2006)	[CP47]	Wood et al. (2013)	[CP60]	Aarne et al. (2018)	
[CP22]	Stotts et al. (2003)	[CP35]	Williams et al. (2006)	[CP48]	Mitchley et al. (2014)	[CP61]	Celepkolu & Boyer (2018)	
[CP23]	Thomas et al. (2003)	[CP36]	Carver et al. (2007)	[CP49]	Ahmad et al. (2015)	[CP62]	Chen & Rea (2018)	
[CP24]	Williams et al. (2003)	[CP37]	Van Toll III et al. (2007)	[CP50]	Kongcharoen & Hwang (2015)	[CP63]	Smith et al. (2018)	
[CP25]	Hanks et al. (2004)	[CP38]	Madeyski (2008)	[CP51]	O'Donnell et al. (2015)	[CP64]	Gold-Veerkamp et al. (2019)	
[CP26]	Melnik & Maurer (2004)	[CP39]	Sison (2009)	[CP52]	Urai et al. (2015)	[CP65]	Jarratt et al. (2019)	
[CP27]	VanDeGrift (2004)	[CP40]	Stapel et al. (2010)	[CP53]	Al-Jarrah & Pontelli (2016)	[CP66]	Ying et al. (2019)	
[CP28]	Bellini et al. (2005)	[CP41]	Lai and Xin (2011)	[CP54]	Isong et al. (2016)	[CP67]	Bowman et al. (2019)	
[CP29]	Layman et al. (2005)	[CP42]	Martínez et al. (2011)	[CP55]	McChesney 2016	[CP68]	Kaurkuttal et al. (2019)	

13.4.1 What are the Measures of Interest to Analyze Effectiveness of Pair Programming in Computer Science Education? [RQ-1]

This question aims to identify the measures of interest used in the articles to examine the effects of PP. For example, one research paper studies PP's effect on developers' productivity. Here developer productivity is a measure of interest. Several metrics are used to measure the effectiveness of PP in computing education including quality of work and exam scores. For example, difference in quality of work of students working in pairs versus solo.

To answer RQ-1, Figure 13.1 presents the measures of interest which are used for studying PP in higher education. We identify and categorize, from the literature review, six measures of interest. The details are given below:

- **Students' performance:** Students are evaluated on the basis of their performance in examinations i.e. final exam scores (Williams et al., 2003; Tsompanoudi et al., 2016; Erdei et al., 2017), or programming lab scores/grades for projects/lab assignments (Bowman et al., 2019). Homework scores are also used for measuring performance in (Bowman et al., 2019). Using final exam scores is the most preferred metric for obtaining data on students' performance as a final exam is conducted in a controlled setting, and is, therefore, more reliable. Moreover, final exams are independent, and scores indicate one's own effort (mitigating the free loader effect). Some studies (Layman et al., 2005) have used midterm exam grades, GPA, and SAT scores as well. Pair compatibility is a major concern to see the effectiveness of PP. Sfetsos et al. (2009) measured pair performance to judge pair compatibility using communication, velocity, design correctness, and passed acceptance tests as the metrics. Whereas, Williams et al. (2006) relied on a peer-evaluation survey to gauge pair compatibility.
- **Learning/understanding:** PP helps in sharing knowledge. Bellini et al. (2005) measured knowledge sharing using two questionnaires – entry and exit, which tested students' knowledge about the maintenance activity of the system used in the experiment. Mendes et al. (2006) also used learning outcome as a dependent variable, and measured the amount of material learnt using grades/marks obtained in assignments, test, and exams. Jones & Fleming (2013) studied 14-hour video recording of a PP experiment using an iterative open coding process to understand the knowledge sharing. In DPP, number of messages exchanged by the partners can also indicate communication, and thus knowledge sharing (Xinogalos et al., 2017).
- **Productivity:** To measure efficiency of PP, researchers measure the time used to finish assignments, or the amount of work accomplished during the experiment. McDowell et al. (2003) and Isong et al. (2016) noted about the time students spent on developing program for the assigned task. Melis et al. (2006) counted *total working days* and *lines of code (KLOCs)* to estimate effort,

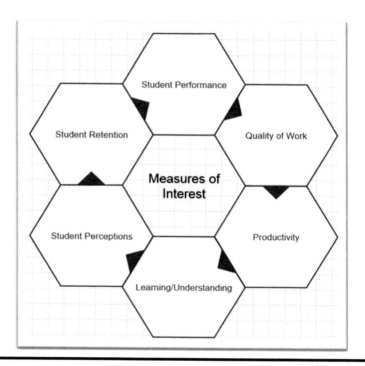

Figure 13.1   Details of measures of interest explored in research studies.

and *released user stories* to estimate the work done. Sison (2009) uses *LOC/person-hours* to measure productivity.

■ ***Quality of work:*** Quality of student work is decided on the basis of grades assigned by instructor for lab work in (Gehringer, 2003; Ahmad et al., 2015), and for homework in (Kongcharoen & Hwang, 2015). McDowell et al. (2003) analyzed functionality and coding styles of students' programs, to evaluate quality of work, submitted as lab assignment for a relatively complex problem. *Defect density (Defects/KLOC)* is also used to measure code quality (Melis et al., 2006; Müller, 2007; Sison, 2009; Rong et al., 2018). Madeyski (2008) focused on thoroughness and fault detection effectiveness of unit test suites to evaluate the effectiveness of PP.

■ ***Student retention:*** It measures students completing the course (i.e. not dropping the course) or persistence of students in Computer Science related majors in future (Williams et al., 2003; Carver et al., 2007). Some also used lab attendance (O'Donnell et al., 2015; Aarne et al., 2018; Jarratt et al., 2019) to measure student retention in class, or their success rate (Williams et al., 2003) in the present course (a grade of "C" or higher).

■ ***Students' perceptions:*** Students are surveyed/interviewed with open-ended/closed questions for understanding their attitude toward PP, and experience of pairing (Melnik & Maurer, 2004; Simon & Hanks, 2008; Ying et al., 2019).

Observations of instructors/lab assistants are also used for understanding students' perceptions (Nawahdah & Taji, 2016). Celepkolu & Boyer (2018) made their students write reflection essays at the end of term to explain their learning process, problem solving approaches, and course content. They used sentiment analysis and thematic analysis to understand students' attitudes, students' reflections on pair programming. Chen & Rea (2018) measured students' attitudes using the Attitudes Towards Mathematics Inventory (ATMI) instrument. Kaurkuttal el al. (2019) analyzed students' perceptions from the codes developed from transcriptions of Videos/Audio recordings of students during the experiment, as well as from their post-interviews.

Although majority of the studies report that students enjoyed the PP practice, many highlighted the negative issues students raised such as lack of concentration due to noise in the lab. PP may affect students' confidence, communication, and satisfaction with their work. Students feel PP improves task efficiency as they are less likely to "get stuck", and working together with peers helps to improve their exam scores as well as their social skills. Negative feedback is mostly the outcome of incompatible pairings, but some noted that individual's personality type, gender, skill level, and learning style can also make one feel bad about the practice.

All the measures that help students to improve their score or gain a skill are here referred as student achievements. Student achievements, after experiencing PP, are in terms of better performance in exams/programming assignments, gaining knowledge/skills, writing good quality code, productivity, interpersonal skills including communication, and completing a course.

13.4.2 What is the Effect of Pair Programming on These Measures of Interest? [RQ-2]

This question analyzes the effect of PP on the measure of interest. As an example, when someone says, "pair programming improves code quality". Here code quality is the measure of interest, and PP has a positive effect as it improves code quality.

It is noted that understanding effectiveness of the PP practice in Computer Science education has been the main area of research in this domain. There is no surprise that, in general, PP is found to have positive effect on various measures of interest with students' performance being the most popular (37% studies) among all (see Table 13.5). Majority of the studies investigated students' performance, and notably many of them have also concluded contradictory results. Nawahdah & Taji (2016) shown negative effect of the PP practice on students' performance. Several studies show mixed effect of the PP practice on students' performance like Radermacher & Walia (2011) observed significant improvement in the grades of only academically weaker students. Equally good number of studies concluded that PP has no effect on students' performance. Many times it happened due to small sample size.

Table 13.5   Measuring Student Achievements and Effect of PP on These Measures

Measures of Interest	Total Papers	Positive Effect	Negative Effect	Mixed Effect	No Significant Effect
Students' performance	26	[JP01] [JP02] [JP12] [JP16] [CP24][CP34] [CP41] [CP50] [CP51] [CP54] [CP63]	[CP56]	[JP09] [JP11] [JP13] [CP20] [CP43] [CP55] [CP62]	[JP14] [CP17] [CP18] [CP21] [CP23] [CP51] [CP60]
Learning/ Understanding	7	[JP09] [JP16] [CP19] [CP43] [CP50]	-	[CP28]	[CP44]
Quality of work	15	[JP01] [JP02] [JP04] [CP20] [CP24] [CP33] [CP42] [CP49] [CP54] [CP56]	-	[CP25] [CP39]	[JP03] [JP14] [CP38]
Productivity	10	[CP43] [CP46] [CP50] [CP53] [CP54]	[CP33]	[JP14] [CP39]	[JP04] [CP20]

Clearly, several empirical studies provide strong support for the conclusion that PP improves quality of work. However, Müller (2007) observed that a student pair lacking requisite skills for solving a problem cannot write better program than a solo programmer, and conclude that PP is better for solving recurring problems whose solutions students already know. In fact, PP can improve quality of work when first year students pair on a relatively complex program (Sison, 2009). However, making senior students pair program simple problems will not yield any positive effect of the PP on quality of work. Similarly, results reported in Hanks et al. (2004) are also inconsistent and contradict previous findings that pairing produces good quality programs.

Table 13.5 shows that PP is helpful in improving productivity as well. However, some studies contradict these findings. Rong et al. (2018) used the time to task completion for measuring productivity and found that there is no significant difference between paired students and solo students as far as their productivity is concerned. Using LOC produced as the metric for measuring productivity, Bipp et al. (2008) also didn't find any significant difference in both the groups – pair v/s solo.

Table 13.6   Effectiveness of PP for Student Retention

Measures of Interest	Total Papers	Positive Effect	Negative Effect	Mixed Effect	No Significant Effect
Attendance	3	[CP56] [CP66]	-	[CP51]	-
Participation	2	[CP20] [CP21]	-	-	-
Course completion	4	[JP02] [JP09] [CP24]	-	[CP21]	-
Opting/ continuing for next course	4	[CP24] [CP46]	-	[JP09]	[CP36]

Besides investigating the effects of PP on students' performance and productivity, researchers also study the efficacy of PP to increase student retention in Computer Science courses (see Table 13.6). Student retention is measured using several metrics as mentioned in Section 13.4.1. In (O'Donnell et al., 2015), the authors found no evidence for a difference in attendance levels between students from the paired and solo groups. Notably, the effects of PP on attendance were significant for academically weaker, and not for stronger students. Braught et al. (2011) found that successful completion of course was interplay of PP and student's SAT score. So student retention cannot be fully attributed to the PP practice. Carver et al. (2007) reported that results were positive but not significant due to small sample size.

13.4.3 *What are Students' Perceptions of Using Pair Programming? [RQ-3]*

Students' perceptions are also used as an objective of investigation in a large number of research papers. How do students perceive or interpret PP? Do they look at PP as a savior for understanding, learning course work from friends/peers in an enjoyable way or get dreadful while sharing time/skills with others, or people from other gender? Students' perception is a student's knowledge of the benefits and difficulties of peer programming. Good experiences (e.g. compatible pairs) with the practice of PP may often lead to affirmative views of the approach. It also depends upon gender, and learning style of an individual.

This research question focuses on students' perceptions as they experience PP practice in their courses. To answer, this research question, Table 13.7 presents the students' perceptions explored in the related PP studies. The common findings in this regard are that majority of the students have positive perceptions of the PP

Table 13.7 Students' Perceptions of PP

Measures of Interest		Total Papers	Positive	Mixed	No Effect
Attitude		17	[JP02] [CP21] [CP24] [CP43] [CP54] [CP56] [CP57] [CP61] [CP64]	[CP23] [CP29] [CP34] [CP51] [CP60] [CP65]	[P13]
Productivity		10	[JP11] [CP26] [CP27] [CP30] [CP32] [CP42] [CP61] [CP67]	[JP06]	[CP20]
Performance		1	[CP57]	-	-
Quality of work		7	[JP02] [CP26] [CP29] [CP30] [CP42] [CP41] [CP47]	-	-
Learning/ understanding		17	[JP06] [CP26] [CP27] [CP30] [CP29] [CP37] [CP42] [CP43] [CP45] [CP51] [CP54] [CP57] [CP61] [CP64] [CP67]	[CP65] [CP66]	
Career skills		3	[CP30] [CP57] [CP66]	-	-
Team skills (e.g. communication, cooperation)		4	[JP06] [CP17] [CP26] [CP64]	-	-
Affective skills	confidence	9	[JP09] [CP25] [CP27] [CP30] [CP47] [CP50]	[CP65]	[JP13] [CP29]
	satisfaction	6	[CP23] [CP25] [P53] [CP61] [CP55] [CP56]	-	-
	motivation	4	[JP02] [JP06] [CP50] [CP64]	-	-
Interpersonal skills		3	[JP02] [CP56] [CP61]	-	-

experience. The students felt that the PP practice increases productivity, improves quality of work, and learning. PP is perceived to be close to the work environment, and hence useful for better employability.

No doubt PP is found to be very successful in improving students' performance, and retaining students in Computer Science, but it is important to consider that all students don't benefit from this approach in the same way. Some students dislike working with others in the PP practice. Some students also had negative experiences such as felt frustrated and dissatisfied with the practice (Braught et al., 2011). There has to be some noise in the lab due to the PP practice as partners discuss their problems and strategies, but some students complain that they cannot concentrate in such a noisy environment (Isong et al., 2016). Similarly, students in the experiment conducted by Gold-Veerkamp et al.(2019) felt negative experience of PP. Hanks (2006) noted that instructors may also influence students' attitude as indicated by one section of students in his experiment.

What are the major reasons of students having negative view of the practice? Here is a list of some concerns reported in the literature:

- ■ *Personality traits:* Students characterized as introverts dislike the practice more. They are more comfortable working alone.
- ■ *Learning style:* Students who disliked pairing were the ones identified as significantly stronger toward the reflective end of the active-reflective learner scale (first dimension of the popular Felder-Silverman scale of learning styles). Learning style of reflective learners is to think solutions on their own, whereas active learners prefer learning by trying things out and working with others.
- ■ *Pair dynamics:* Pair composition plays an important role in students disliking the practice. Lack of compatibility with partner often spurs bad experience of PP practice in students.
  - o *Free-loader problem:* For example, pairing with a free-loader (who is not actively engaged in the task) as the programming partner seems wastage of time as it is believed that a non-contributing partner slows down the speed of work by distracting the working partner, and increases workload.
  - o *Student conflicts on choosing a solution:* We know that there are multiple ways of solving a problem. Some students don't appreciate partner's point of view and insist on using their own logic (Martinez et al., 2011; McChesney, 2016). Therefore, the partner may feel disassociated with the solution and becomes disinterested in the practice.
  - o *Coordination problems:* Some students find it difficult to coordinate with other students irrespective of their skill levels. It has been observed that students with same skill levels also sometimes fail to coordinate like not agreeing on a solution, or doing work in one's own way, or being apprehensive of one's own as well as of partner's abilities. Another coordination problem is related to failure of agreeing on a common time schedule in their academic and extracurricular activities, when there are no fixed lab

hours, for performing assignments using the PP practice (VanDeGrift, 2004; Simon & Hanks, 2008).

o *Different skill levels:* The studies conducting PP experiments also report conflicts between students' knowledge levels as one of the reasons of disliking the practice. A weaker student feels left behind when more skilled partner dominates the scene, and doesn't bother to explain and take along the weaker partner. A skilled partner feels burdened for the weaker partner being hardly of any value in solving the problem (O'Donnell et al., 2015). Therefore, academic compatibility may be a precursor for students' positive attitude toward PP.

o *Gender issues:* Some women feel lack of confidence when they code in front of men. They believe that men don't allow their female partners to express their ideas, what to talk about trying something new in their presence. In some conservative societies, social customs repel forming mixed gender pairs.

■ ***Instructors' role:*** Instructor's attitude may also influence students' attitude. If an instructor is not himself convinced with the approach, he may fail to sensitize students about its benefits.

■ ***Technical issues:*** Some students are not comfortable with sharing computing devices (Celepkolu & Boyer, 2018). In some societies, students from different genders feel awful when seated in proximity for sharing computer monitor and keyboard. The issues related to technology also surface in DPP where students feel lack of an incessant Internet connection. Besides unexpected issues in service provider's technical infrastructure, inefficient student owned infrastructure also spoils programming experience in DPP.

When a partnership does not function successfully, the pair is less productive and may become disinterested in the practice (Ying et al., 2019). Pair formation on the basis of student friendships is expected to produce positive student experience, but whether such pair formation brings other benefits of PP such as performance and productivity remains another concern.

13.4.4 What are the Factors that Determine Effectiveness of PP? [RQ-4]

Educators are still susceptible of the practice when confronted with a choice of using PP in classroom settings. Using PP as a pedagogical aide entails considering the reasons why PP might or might not be effective in a particular context. There is need to identify conditions that are required for PP to be useful e.g. Pair Compatibility. Moreover, PP as a pedagogical tool can be effective, if it appeals to all students irrespective of their differences of learning style, personality type, gender, or skill levels.

The previous research questions (RQ-2 and RQ-3) analyzed if students working in pairs v/s solo had an effect on their performance in exam/tests, quality of work, learning, and attitudes. This research question (RQ-4) answers whether such an effect differed depending upon personality, learning style, gender, skills, or other factors. Here we address studies which find answers to questions like "Did women who pair programmed successfully complete the course more often than men, or vice versa?" This question is posed to better understand the affect that various factors like learning style, personality type, gender, or skill levels may have in making PP effective or ineffective. For example, some studies report that PP is more effective for women than men (McDowell et al., 2006; Ahmad et al., 2015), and also for weaker students (O'Donnell et al., 2015).

As already observed in RQ-3, some students have negative perception of PP and dislike the practice. Human issues like skills, personality, and gender can play an important role in the success/failure of PP practice. RQ-4 explores the empirical support for conclusions in this regard. Personality type, for example, did not show any significant effect on PP effectiveness in the research experiments by Layman (2006) and Choi et al. (2009). Williams et al. (2006) found support for predicting compatible pairs when partners with sensing-intuition dimension of personality types joined. These studies use Myers–Briggs[1] personality scale for profiling students on the basis of their personality types. Sfetsos et al. (2009) analyzed the impact of developer personalities/temperaments (using the Keirsey Temperament Sorter) on pair effectiveness, and observed that heterogeneous pairs performed better than homogenous pairs. Radermacher et al.(2012) used the concept of mental model consistency (MMC) to form pair, and found it as a predictor of performance in PP. Salleh et al. (2014) used the Five Factor Model (FFM) of personality traits, which has five broad personality traits – Conscientiousness, Extraversion, Agreeableness, Neuroticism, Openness to experience, and analyzed the three factors (Conscientiousness, Neuroticism, Openness to experience). Their research shows that Openness to experience had a significant impact on pairs' performance, whereas Conscientiousness and Neuroticism didn't have any.

Researchers have explored various pair dynamics (see Table 13.8). Instructors can pair up the students using random pairing scheme in which students' personality, skills, or gender are not taken into consideration. Random pairing leads to incompatible partners and raises conflicts among student pairs (Nagappan et al., 2003). Students also feel concerned that both the partners will be evaluated at same level for the activity, so difference in knowledge level demotivates skilled students from pairing with less skilled. This may be the reason, with few exceptions, that most students prefer to work with a similar skilled partner. Xinogalos et al. (2017) observed that students prefer friendship relations as programming partners when given an option.

Academic skills (such as GPA, SAT scores) are not good predictors of compatible pairs (Williams et al., 2006). Some researchers observed that student pairs with perceived similar skills are more compatible than with actual similar skills

Table 13.8 Some Methods for Student Pairing in an Education Setting

Pairing Method	Article-ID
Random	[CP20] [CP21] [CP65]
Self-selected	[JP02] [CP38] [CP57]
Instructor-assigned	[JP05]
Programming attitude	[CP23]
Gender	[JP07] [P15] [CP68]
Skill	[JP01] [JP05] [JP12] [CP35] [CP47] [CP64] [CP67]
Personality-traits	[JP07] [JP08] [JP11] [CP35]
Mental model consistency	[CP44]
Learning-style	[CP35]
Self-esteem	[CP23] [CP35]
Work-ethics	[CP35]
Organizational-skills	[CP35]

(Williams et al., 2006; Radermacher & Walia, 2011). In a recent study, Gold-Veerkamp et al. (2019) considered different pair compositions in PP on the basis of their knowledge skills, and formed homogenous (weak/weak, strong/strong), and heterogeneous (weak/strong) pairs. The study reports that PP is equally effective for all kinds of pair compositions – homogenous as well as heterogeneous. Bowman et al. (2019) found that prior experience of a partner doesn't make any difference in making PP more effective, rather it demotivates the less experienced partner. Thomas et al. (2003) put students, on the basis of their self-reported programming experience before university, into three categories – Code-Warrior, Middle, and Code-a-Phobes arranged from high to low. Their self-perception survey results show that students with considerable self-confidence in their programming skills, when paired with low-confidence students, did not enjoy PP as much as the other students. But pairs with similar self-confidence levels produced their best. Bellini et al. (2005) tested effectiveness of PP when partners have different educational backgrounds (scientific versus non-scientific). Using knowledge sharing as a dependent variable, the study reports that PP is beneficial for pairs with same educational background, and not for the mixed pairs.

Several have studied the role of gender in PP. McDowell et al. (2006) reported no significant gender difference in student's performance after PP exercise. However, men working in pairs enjoyed more than the ones working solo. But there was no significant difference in enjoyment for women in pairs versus solo. PP induced significantly more confidence in men than women. However, women working in pairs were more confident than the ones working solo. PP seems to be more conducive to same gender pairs (Choi et al., 2009). As per Choi et al., students in the experiment apparently benefitted from the practice more as indicated by gains in communication, satisfaction, and compatibility. Members from the same gender enjoy the activity and appear more comfortable with each other in the physical setup of a collocated PP experiment. However, such pairs are less confident of their programs. Intuitively, learning should be gender-inclusive. But it has been noted that pairs which are more democratic belong to the same-gender. In mixed-gender pairing, women particularly don't benefit (Jarratt et al., 2019), and they should be paired with women for better results (Choi et al., 2009). Jarratt et al. (2019) reported that with mixed-gender pairs, lab attendance is good, confidence in the finished product is good, but productivity is less. Kaurkuttal et al. (2019) analyzed pairs in a DPP setup to understand gender effects. They report that same gender pairs were more democratic than mixed-gender pairs in driver/navigator roles as well as in task management. Same gender pairs were also more compatible, and more confident. A female partner is presumed to benefit both male and female students (Jarratt et al., 2019).

Student pairs may be fixed for a whole semester, or may change frequently through the semester as in (Williams et al., 2006; Wood et al., 2013). There is no comparison of both the strategies. Jones & Fleming (2013) call the frequent changing of partners as promiscuous pairing, and recommend that pairing with many partners gives better exposure, and hence increases programmer expertise.

Several papers support the hypothesis that women find PP more effective than men (McDowell et al., 2006). Aarne et al. (2018) noted statistically significant differences between men and women students in PP activity e.g. women attended more PP labs. Recently, Ying et al. (2019) reported, on surveying 104 students, that there were hardly any gender differences in students' sentiments about their experience with PP.

Experiments exploring effect of learning styles indicate that learning style has no significant impact on effectiveness of PP (Layman, 2006; Williams et al., 2006). Further, PP is more effective for academically weaker students rather than the ones with good skills (O'Donnell et al., 2015). Therefore, it shows that skill level may affect the effectiveness of PP. Problem complexity also plays an important role. Senior students will not gain much from solving simple problems using PP approach. For PP to be effective in such cases, programming task has to be reasonably complex.

Though PP approach is mostly analyzed for teaching of programming subjects, but some researchers experimented with other subjects such as Basic Microsoft Excel (Chen & Rea, 2018) and Computer Networks (Kongcharoen & Hwang, 2015). Chen & Rea (2018) also investigated other subjects such as Microsoft Office Suite, Basic Web Creation, Advanced Microsoft Access and Excel, and Data mining. The results indicate PP practice is equally effective for different subjects.

PP may also help Computer Science educators to reduce their workload by utilizing the PP practice as a pedagogical tool. Instructor feedback in some of the studies shows positive perception among the faculty about PP as it helps in reducing workload in lab (Nagappan et al., 2003; Kongcharoen & Hwang, 2015), and in submission evaluation (Rong et al., 2018). However, Erdei et al. (2017) has interesting results when they experimented using a teaching assistant for one cohort, and PP as a teaching aid for another cohort. They found that there was no significant difference in students' performance in both the cohorts for the first two programming examinations in the semester. But in the third examination, the cohort using PP as a teaching aid performed worse. Hence, we believe that PP practice can be useful in teaching large classes when they are exploring introductory concepts, but it may not be appropriate for teaching advanced concepts.

13.4.5 Is Pair Programming Methodology Beneficial for Students' Teaming with Geographically Dispersed Partners? [RQ-5]

With the significant penetration of the Internet technology, digital education is experiencing explosive growth. Remote students can form virtual teams to work on team projects and get benefits of PP.

Distributed pair programming (DPP), also known as remote or virtual pair programming, has been known to be comparable to co-located pair programming in regards to students' performance, code quality, and productivity (see Table 13.9). In the research literature on DPP in education, empirical studies compare DPP against collocated solo Zacharis (2011), and distributed solo

Table 13.9 Effectiveness and Student Perceptions of Distributed Pair Programming

Measures of Interest	Sub-Items	Positive	No Significant Effect	Mixed Effect
Students' performance	Students' grades	[JP12]	[JP05] [JP10]	[CP64]
	Code quality	[JP10] [CP22]	-	-
	Productivity	[JP10] [CP22]	[CP22] [CP58]	-
Pair compatibility		[CP59]	-	-
Scheduling issues		-	-	[JP05]
Student perceptions		[JP12] [P15] [CP31]	[CP64]	[CP59]

(Stotts et al., 2003) teams. Baheti et al. (2002) conducted first experiment to measure efficiency of DPP in education. In the experiment reported in Hanks (2008), students in the experimental group had same levels of performance, measured using programming assignments and final exam scores, as collocated students. Zacharis (2011) investigated the effectiveness of VPP on students' performance and satisfaction in an introductory programming course. Students' performance of pairs, measured using programming assignment grades, was not significantly different from the solo group. Code quality, measured using defect count, was significantly better for pairs. However, code productivity was not different for both the groups, pair and solo. Students' perception of the DPP experience was, however, highly positive. Tsompanoudi et al. (2019) found that the individual with more programming experience and confidence secured higher assignment grades while the individual with lower perceived pair compatibility secured higher exam grades. In Hanks (2008) results show preference for DPP commensurate with confidence. Tsompanoudi et al.(2019) and Kaurkuttal et al. (2019) observed that women preferred collocated PP, and men were more okay with DPP.

DPP extends the practice of PP by allowing the partners to pair unfettered by physical location. Research suggests that DPP matches Collocated PP (CPP) in benefits and in addition gives collaborators freedom of time and location.

However, DPP remains more challenging as students need appropriate communication and collaboration infrastructure (a shared editor, and a communication tool being the most basic requirements) so that they can schedule at any time and work effectively from their convenient location. Distributed pairs need support for communication and collaboration as the physical environment of the CPP provides some of the activities (such as body language to understand partner actions conveniently, quickly, and naturally) that cannot be replicated in an online environment without much commotion. Hanks (2008) recommend use of a gesturing pointer (e.g. a pointing index finger) in a DPP tool (VNC4DPP) to draw driver's attention toward navigator's feedback. Students' developers used the pointer regularly, and found it useful. However, the tool usage didn't have any significant effect on students' achievements. Urai et al. (2015) highlighted the need for support functions in DPP environments for productivity gains as they observed in an experiment that pairs spent a lot of time in communicating their intent. In CPP, pairs can use vocal communication as well as gestures and pointers. Pairs in DPP need support functions that can help to easily share code and outputs. A DPP environment should have functions like text chats and annotated code. Tsompanoudi et al. (2016) proposed using collaboration scripts to guide students to perform activities in a well-defined manner for an effective experience. However, a pilot run of the proposed framework showed that students like the DPP setup, but didn't like the scripted guidance for taking turns while programming. Such an arrangement may be helpful for novice programmers. Their next experiment gave relaxation on role switching by leaving it to some extent to the pair to decide (and not the script), and results showed that both the approaches of role assignment were equally effective measured using students' contribution rates. Al-Jarrah & Pontelli (2016) evaluated the *AliCe-ViLlagE* platform, a collaborative VPP environment based on *ALICE*, and found it good for students' productivity and satisfaction. Tsompanoudi et al. (2016) developed an Eclipse plugin, SCEPPSys (Scripted Collaboration in an Educational Pair Programming System), to support DPP in an educational setting, and it worked well for student collaboration and balanced participation in DPP. However, Zacharis (2011) carried out the DPP experiment without using any dedicated collaboration tool for DPP, and students reported highly positive attitude toward the practice. Their students used Windows based tools like NetMeeting and the Remote Desktop Sharing, and free VoIP applications like Skype for real-time communication.

13.5 Threats to Validity

We follow the guidelines by Wohlin et al. (2000) to discuss the threats to validity of the results presented here.

Construct validity refers to the relation between the theory and the observations. Researcher bias is mitigated by following a systematic approach of literature review as per Kitchenham & Charters (2007). Research questions are defined in alignment with the goals of the review. Moreover, the search string was refined after performing pilot searches. Published articles were retrieved from well-managed, most trusted, and widely used sources of literature.

Internal validity refers to the causal relationship between treatment and outcome. While selecting articles, we followed well-defined inclusion/exclusion criteria. A rigorous searching and selection process helped to mitigate the threat of missing out a related article.

Conclusion validity is concerned with the ability to draw the correct conclusions from the study. Threats in this case can be mitigated by following a rigorous classification scheme. As discussed in Section 13.3, the coding process using thematic analysis was cross validated to deal with the validity of the results.

13.6 Issues and Future Work

Despite the extensive research literature on PP in the field of CS education, there are still a series of issues that raise open problems for future work in this field. Previous studies have identified issues and open problems. But only a few researchers discussed the solutions.

As per the in-depth analysis, key concerns include measures of interest, pair composition, coordination and communication, problem complexity, external motivations, instructor attitude, sample size, and logistics.

Just focusing on performance in exams as an indicator of PP effectiveness may be misleading. Some students are result oriented, and they can go to any extent (read: study harder or take private tuitions near the exams) to score high in exams. Moreover, question paper style in the exam (recall versus conceptual questions) may also affect exam performance. No study provides the details of the examination, which we believe is very important before attributing better performance to PP. Recently, Chen & Rea (2018) identified "enhanced sense of accomplishment" as a factor for predicting improvement in students' scores in a PP experiment. Therefore, internal validity of PP experiments measuring students' performance is of major concern as some other factors may have caused the effect. Students' performance in a supervised PP lab may be different from an unsupervised one. How do students' work in pairs when no one is monitoring them? It has been observed that students follow the PP implementation guidelines only partially (Gold-Veerkamp et al., 2019).

Student retention using "completing a course" as a measure may not be attributed to PP when course credits are fixed and students' have very few options to pick other elective courses. Moreover, programming subjects are the core of Computer Science. So it is important to keep in mind the external motivations.

There is also need to focus on high achieving students to find how they can benefit from the practice. More research into students' self-efficacy, and attitudes toward PP is required to reach a consensus as existing studies are contradictory (Thomas et al., 2003; Hanks, 2006; Aarne et al., 2018).

Furthermore, just one or two experiments are not enough to assess the effectiveness. Some longitudinal studies using PP have observed different results of students' performance in the ensuing experiments (Gehringer, 2003; Bowman et al., 2019). Though, Gold-Veerkamp et al.(2019) and McChesney (2016) didn't see any change in attitudes i.e. students liked the practice after repeated use in different semesters. There is need to repeat these experiments in different contexts to understand how do students with different profiles acclimatize to the practice.

Programming tasks given to the pairing v/s solo may also impact research results. Most of the existing studies presented same task to both the cohorts – experimental and control. Future studies could seek to find pair performance by varying the task across the cohorts, and across the different experiments. Programming task should be of appropriate complexity to demand skills from both the partners.

Pairing dynamics and compatibility (Williams et al., 2006; Kaurkuttal et al., 2019) is one of the most widely discussed topics in the research literature. Demographic differences in students working as pair may also affect their experience with the practice. Pair compatibility research should also explore cultural differences, and cognitive inequalities. Mixed-gender pairs may have issues of ego, sense of superiority (in some cultures), and competitiveness. Moreover, future work can explore students' aims or aspirations as a criterion while forming pairs. Students aiming to be professional programmers are keener to learn the skill, while some are less motivated, less engaged, and just want to pass the course.

While taking into consideration academic background for forming pairs, we should understand that students with same score have different levels of conceptual clarity. Therefore, academic score may also not be a true indicator of one's knowledge of the subject as learning styles (rote learning versus learning by doing/understanding), and question paper format during the exam may be some of the confounding factors. Therefore, lack of context details while reporting PP experiments for publication obscures the validity of the results.

Another concern could be how student pairs share responsibility between themselves for success or failure of a program. Pair composition experiments should also devise mechanisms to decide division of labor between the pair to reduce free-loader problem. For example, several studies depended upon partner feedback, through an online system, to detect pairs with unequal participation (Gehringer, 2003; Layman et al., 2005; Williams et al., 2006). Gehringer (2003) also recommend frequent pair switching to manage free-loaders in a class. Tsompanoudi et al. (2016) used collaboration scripts to address the concern of unequal participation in a PP experiment. Later, the ratio of the typed code to the total code helps to decide individual contribution.

Student pairs may have to manage several conflicts including different time schedule, ideas, culture, and language (McChesney, 2016). How can they learn to manage these conflicts? As employability or giving real world job experience is also sometimes mentioned as benefits of PP, students should learn how to manage personal differences in real life, and be productive.

Finding effective modes of coordination, communication, collaboration overall in PP, and DPP in particular is also an important challenge. In the existing studies, several tools are demonstrated. They need to be evaluated by other researchers and educators for understanding their strengths and weaknesses, and using them in practice. Furthermore, evaluating the impact of students in a pair with different spoken languages on the efficiency of their communication and collaboration in PP is also important.

Most of the experiments in the existing literature are conducted by instructors who are themselves interested in the PP practice as researchers. Therefore, investigating instructor attitude, in general, is an interesting and important challenge, because some instructors don't allow PP to guard against free-loaders. In addition to student attitudes toward PP, research should also investigate instructor's attitude toward the practice. Future work should also explore the interaction between students' and instructors' attitudes for understand PP effectiveness.

In a PP environment, even design of the physical space is important. Bettin et al. (2019) proposed to value physical (lab) space design while analyzing the effectiveness of PP. What is more ideal "one computer, one mouse, one keyboard" or "one computer, two monitors, two mice, two keyboards"? Investigating the context, e.g. a different desk layout, screen size, or lab setup may provide new insights or may impact research results e.g. student perceptions.

Last but not the least, a lot of studies use small sample size, and therefore cannot discuss the significance of the results. For example, Tsompanoudi et al. (2019) and

Kaurkuttal et al. (2019) experimented with only six pairs each, which is not enough to make a generalizable statement. A follow-up research agenda can be to replicate these studies on a large sample size, and add more empirical support. Another factor that needs more experimentation is the longevity of the PP experiments. For example, results of experiments running for multiple semesters (Williams et al., 2000; McChesney, 2016) cannot be seen in the same light as a single 50-minute class session (Radermacher & Walia, 2011).

13.7 Conclusion

PP is the research field that deals with the use of synergy of two individuals solving the same programming problem at the same time on shared computing resources. As PP practice has supposedly several benefits for practicing as well as student programmers, empirical analysis of the practice to evaluate its efficacy in different contexts is a very important research objective. First research results of its efficacy in the higher education setting occurred in 2000.

This chapter presents a SLR of the PP practice in Computer Science education for university/college students. Searching five electronic databases gives 1744 articles. Finally, as per the inclusion and exclusion criteria, 68 articles were found suitable for further analysis to answer the five research questions. The data extracted from them helped to observe the PP practice from the following aspects: measures of interest in the PP studies, effect of PP on the measures of interest, effectiveness of PP, student perceptions of the PP practice, and using the PP practice in a geographically distributed environment.

The conclusions of the systematic review are drawn as follows: (1) A variety of measures of interest are used in the research field to evaluate efficacy of the PP practice in higher education e.g. student performance, quality of work, learning, productivity, student retention, and student perceptions. Out of these, student performance dominates the studies followed by student perceptions. (2) Effectiveness of PP is not yet a global truth. As many studies report contradictory results, several research opportunities emerge, such as to study large sample size, to make experiment context clear for increasing educator's confidence in the practice and researchers' interest in experiment replication, and to understand instructor attitudes. (3) Finding effective modes of coordination, communication, collaboration overall in PP, and DPP in particular is also an important challenge. We hope this study will help in advancing the field serving as a reference not only for researchers but for educators as well.

Note

1. http://www.humanmetrics.com/cgi-win/JTypes2.asp

References

Aarne, O., Peltola, P., Leinonen, J., & Hellas, A. 2018. A study of pair programming enjoyment and attendance using study motivation and strategy metrics. In Proceedings of the 49th ACM Technical Symposium on Computer Science Education (SIGCSE '18). Association for Computing Machinery, New York, NY, 759–764. DOI: https://doi.org/10.1145/3159450.3159493

Ahmad, M., Abd Razak, A. H., Omar, M., Yasin, A., Romli, R., Abdul Mutalib, A., & Zahari, A. S. 2015. The impact of knowledge management in pair programming on program quality. Abraham et al. (eds.), Pattern Analysis, Intelligent Security and the Internet of Things, Advances in Intelligent Systems and Computing, 159–168. Springer International Publishing Switzerland. DOI: 10.1007/978-3-319-17398-6_15

Al-Jarrah, A., & Pontelli, E. 2016. On the effectiveness of a collaborative virtual pair-programming environment. In: Zaphiris, P., & Ioannou, A. (Eds.). Learning and Collaboration Technologies. Lecture Notes in Computer Science. Springer, Cham. DOI: 10.1007/978-3-319-39483-1_53

Baheti, P., Gehringer, E., & Stotts, D., 2002. Exploring the efficacy of distributed pair programming. Proceedings of the Second XP Universe and First Agile Universe Conference on Extreme Programming and Agile Methods Extreme Programming/Agile Universe, Chicago, IL, Springer-Verlag, Berlin, Heidelberg, 208–220.

Bellini, E., Canfora, G., García, F., Piattini, M., & Visaggio, C. 2005. Pair designing as practice for enforcing and diffusing design knowledge. Journal of Software Maintenance and Evolution: Research and Practice, 17(6), 401–423. John Wiley & Sons. DOI: https://doi.org/10.1002/smr.322

Bettin, B., Ott, L., and Ureel, L. (2019). More Effective Contextualization of CS Education Research: A Pair-Programming Example. In Innovation and Technology in Computer Science Education (ITiCSE '19), July 15–17, 2019, Aberdeen, Scotland Uk. ACM, New York, NY, USA, 7. https://doi.org/10.1145/3304221.3319790

Benadé, T., & Liebenberg, J. 2017. Pair programming as a learning method beyond the context of programming. In Proceedings of the 6th Computer Science Education Research Conference (CSERC '17). Association for Computing Machinery, New York, NY, 48–55. ACM DOI: https://doi.org/10.1145/3162087.3162098

Bipp, T., Lepper, A., & Schmedding, D. 2008. Pair programming in software development teams – an empirical study of its benefits. Information and Software Technology, 50(3), 231–240. DOI: 10.1016/j.infsof.2007.05.006

Bowman, N., Jarratt, L., Culver, K., & Segre, A. 2019. How prior programming experience affects students' pair programming experiences and outcomes. In Proceedings of the 2019 ACM Conference on Innovation and Technology in Computer Science Education (ITiCSE '19). Association for Computing Machinery, New York, NY, 170–175. ACM DOI: https://doi.org/10.1145/3304221.3319781

Braught, G., Wahls, T., & Eby, L. 2011. The case for pair programming in the computer science classroom. ACM Transactions on Computing Education, 11(1), Article 2 (February 2011), 21 pages. DOI: https://doi.org/10.1145/1921607.1921609

Carver, J. C., Henderson, L., He, L., Hodges, J., & Reese, D. 2007. Increased retention of early computer science and software engineering students using pair programming. 20th Conference on Software Engineering Education & Training (CSEET'07). DOI: 10.1109/cseet.2007.29

Celepkolu, M., & Boyer, K. 2018. Thematic analysis of students' reflections on pair programming in CS1. In Proceedings of the 49th ACM Technical Symposium on Computer Science Education (SIGCSE '18). Association for Computing Machinery, New York, NY, 771–776. ACM DOI: https://doi.org/10.1145/3159450.3159516

Chen, K. & Rea, A. (2018). Do Pair Programming Approaches Transcend Coding? Measuring Agile Attitudes in Diverse Information Systems Courses. Journal of Information Systems Education, 29(2), 53–64.

Choi, K., Deek, F., & Im, I. 2009. Pair dynamics in team collaboration. Computers in Human Behavior, 25(4), 844–852. https://doi.org/10.1016/j.chb.2008.09.005

da Silva Estacio, B., & Prikladnicki, R. 2015. Distributed pair programming: a systematic literature review, Information and Software Technology, 63, 1–10.

Erdei, R., Springer, J., & Whittinghill, D. 2017. An impact comparison of two instructional scaffolding strategies employed in our programming laboratories: employment of a supplemental teaching assistant versus employment of the pair programming methodology, 2017 IEEE Frontiers in Education Conference (FIE), Indianapolis, IN, 2017, pp. 1–6. DOI: 10.1109/FIE.2017.8190650

Gehringer, E. 2003. A pair-programming experiment in a non-programming course. In Companion of the 18th Annual ACM SIGPLAN conference on Object-Oriented Programming, Systems, Languages, and Applications (OOPSLA '03). Association for Computing Machinery, New York, NY, USA, 187–190. ACM DOI: https://doi.org/10.1145/949344.949397

Gharoshi, S., & Jensen, C. May, 2017. Integrating collaborative and live coding for distance education. IEEE Computer, pp. 27–35.

Gold-Veerkamp, C., Klopp, M., & Abke, J. 2019. Pair programming as a didactical approach in higher education and its evaluation. 2019 IEEE Global Engineering Education Conference (EDUCON), Dubai, United Arab Emirates, pp. 1055–1062. IEEE DOI: 10.1109/EDUCON.2019.8725150

Hanks, B. 2006. Student attitudes toward pair programming. In Proceedings of the 11th Annual SIGCSE Conference on Innovation and Technology in Computer Science Education (ITICSE '06). Association for Computing Machinery, New York, NY, 113–117. DOI: https://doi.org/10.1145/1140124.1140156 ACM

Hanks, B. 2008. Empirical evaluation of distributed pair programming. International Journal of Human-Computer Studies, 66(7), 530–544. DOI: 10.1016/j.ijhcs.2007.10.003

Hanks, B., Fitzgerald, S., McCauley, R., Murphy, L., & Zander, C. 2011. Pair programming in education: a literature review. Computer Science Education, 21(2), 135–173. DOI: 10.1080/08993408.2011.579808

Hanks, B., McDowell, C., Draper, D., & Krnjajic, M. 2004. Program quality with pair programming in CS1. In Proceedings of the 9th Annual SIGCSE Conference on Innovation and Technology in Computer Science Education (ITiCSE '04). Association for Computing Machinery, New York, NY, USA, 176–180. DOI: https://doi.org/10.1145/1007996.1008043

Hannay, J., Dybå, T., Arisholm, E., & Sjøberg, D. 2009. The effectiveness of pair programming: a meta-analysis. Information and Software Technology. DOI: 10.1016/j.infsof.2009.02.001

Heiberg S., Puus U., Salumaa P., & Seeba A. 2003. Pair-programming effect on developers productivity. In: Marchesi M., Succi G. (Eds.). Extreme Programming and Agile Processes in Software Engineering. XP 2003. Lecture Notes in Computer Science, vol 2675. Springer, Berlin, Heidelberg. DOI: 10.1007/3-540-44870-5_27

Isong, B., Moemi, T., Dladlu, N., Motlhabane, N., Ifeoma, O., & Gasela, N. 2016. Empirical confirmation of pair programming effectiveness in the teaching of computer programming. 2016 International Conference on Computational Science and Computational Intelligence (CSCI). DOI: 10.1109/csci.2016.0060

Janes, A., Russo, B., Zuliani, P., & Succi, G. 2003. An empirical analysis on the discontinuous use of pair programming. In: Marchesi M., Succi G. (Eds.). Extreme Programming and Agile Processes in Software Engineering. XP 2003. Lecture Notes in Computer Science, vol 2675. Springer, Berlin, Heidelberg. DOI: https://doi.org/10.1007/3-540-44870-5_26

Jarratt, L., Bowman, N., Culver, K., & Segre, A. 2019. A large-scale experimental study of gender and pair composition in pair programming. In Proceedings of the 2019 ACM Conference on Innovation and Technology in Computer Science Education (ITiCSE '19). Association for Computing Machinery, New York, NY, 176–181. ACM DOI: https://doi.org/10.1145/3304221.3319782

Jones, D. L., & Fleming, S. D. 2013. What use is a backseat driver? A qualitative investigation of pair programming. 2013 IEEE Symposium on Visual Languages and Human Centric Computing. DOI: 10.1109/vlhcc.2013.6645252

Kaurkuttal, S., Gerstner, K., & Bejarano, A. 2019. Remote pair programming in online CS education: investigating through a gender lens. 2019 IEEE Symposium on Visual Languages and Human-Centric Computing (VL/HCC). pp. 75–85. IEEE DOI: https://doi.org/10.1109/vlhcc.2019.8818790

Kavitha, R., & Ahmed, M. 2015. Knowledge sharing through pair programming in learning environments: an empirical study. Education and Information Technologies, 20(2) (June 2015), 319–333. Kluwer Academic Publishers. DOI: https://doi.org/10.1007/s10639-013-9285-5

Kitchenham, B., & Charters, S. 2007. Guidelines for performing systematic literature reviews in software engineering, technical report EBSE-2007-01, School of Computer Science and Mathematics, Keele University.

Kongcharoen, C., & Hwang, W.-Y. 2015. A study of pair-programming configuration for learning computer networks. 2015 8th International Conference on Ubi-Media Computing (UMEDIA). DOI: 10.1109/umedia.2015.7297488

Lai, H. & Xin, W. 2011. An experimental research of the pair programming in java programming course. Proceeding of the International Conference on e-Education, Entertainment and e-Management. DOI: 10.1109/iceeem.2011.6137800

Layman, L. 2006. Changing students' perceptions: an analysis of the supplementary benefits of collaborative software development. 19th Conference on Software Engineering Education & Training (CSEET '06), Turtle Bay, HI, 2006, pp. 159–166. IEEE DOI: 10.1109/CSEET.2006.10

Layman, L., Williams, L., Osborne, J., Berenson, S., et al. 2005. How and why collaborative software development impacts the software engineering course. In Proceedings Frontiers in Education 35th Annual Conference, Indianapolis, IN, 2005, pp. T4C-T4C. IEEE DOI Bookmark:10.1109/FIE.2005.1611964

Li, Z. Plaue, C., & Kraemer, E. 2013. A spirit of camaraderie: the impact of pair programming on retention. 2013 26th International Conference on Software Engineering Education and Training (CSEE&T). DOI: 10.1109/cseet.2013.6595252

Madeyski, L. 2008. Impact of pair programming on thoroughness and fault detection effectiveness of unit test suites research section. Software Process Improvement and Practice, 13(3), 281–295. Wiley InterScience. DOI: 10.1002/spip.382

Martínez, L., Licea, G., Juárez, J., & Aguilar, L. 2014. Experiences using PSP and XP to support teaching in undergraduate programming courses. Computer Application in Engineering Education, 22(3) (September 2014), 563–569. DOI: https://doi.org/10.1002/cae.20581

McChesney, I. 2016. Three Years of Student Pair Programming – Action Research Insights and Outcomes SIGCSE '16, March 02-05, 2016, Memphis, TN, USA DOI: http://dx.doi.org/10.1145/2839509.2844565

McDowell, C., Hanks, B., & Werner, L. 2003. Experimenting with pair programming in the classroom. In Proceedings of the 8th Annual Conference on Innovation and Technology in Computer Science Education (ITiCSE '03). Association for Computing Machinery, New York, NY, USA, 60–64. DOI: https://doi.org/10.1145/961511.961531

McDowell, C., Werner, L., Bullock, H., & Fernald, J. 2006. Pair programming improves student retention, confidence, and program quality. Communications of the ACM, 49(8) (August 2006), 90–95. ACM DOI: https://doi.org/10.1145/1145287.1145293

Melis, M., Turnu, I., Cau, A., & Concas, G. 2006. Evaluating the impact of test-first programming and pair programming through software process simulation. Software Process: Improvement and Practice, 11(4), 345–360. DOI: 10.1002/spip.286

Melnik G., & Maurer F. 2004. Introducing agile methods: three years of experience. In Proceedings. 30th Euromicro Conference. Rennes, France, 334–341. IEEE DOI: 10.1109/EURMIC.2004.1333388

Mendes, E., Al-Fakhri, L., & Luxton-Reilly, A. 2006. A replicated experiment of pair-programming in a 2nd-year software development and design computer science course. In Proceedings of the 11th Annual SIGCSE Conference on Innovation and Technology in Computer Science Education (ITICSE '06). Association for Computing Machinery, New York, NY, 108–112. DOI: https://doi.org/10.1145/1140124.1140155

Mitchley M., Dominguez-Whitehead Y., & Liccardo S. 2014. Pair programming, confidence and gender considerations at a South African University. In: Thege B., Popescu-Willigmann S., Pioch R., & Badri-Höher S. (Eds.). Paths to Career and Success for Women in Science. Springer VS, Wiesbaden. DOI: 10.1007/978-3-658-04061-1_8

Müller, M. 2007. Do programmer pairs make different mistakes than solo programmers? Journal of System Software, 80(9) (September 2007), 1460–1471. Science Direct. DOI: https://doi.org/10.1016/j.jss.2006.10.032

Nagappan, N., Williams, L., Ferzli, M., Wiebe, E., et al. 2003. Improving the CS1 experience with pair programming. In Proceedings of the 34th SIGCSE Technical Symposium on Computer Science Education (SIGCSE '03). Association for Computing Machinery, New York, NY, USA, 359–362. DOI: https://doi.org/10.1145/611892.612006

Nawahdah, M., & Taji, D. 2016. Investigating students' behavior and code quality when applying pair-programming as a teaching technique in a Middle Eastern society. 2016 IEEE Global Engineering Education Conference (EDUCON), Abu Dhabi, 2016, pp. 32–39. IEEE DOI: 10.1109/EDUCON.2016.7474527

Nosek, J. 1998. The case for collaborative programming. Communications of the ACM, 41, 105–108.

O'Donnell, C., Buckley, J., Mahdi, A., Nelson, J., & English, M. 2015. Evaluating pair-programming for non-computer science major students. In Proceedings of the 46th ACM Technical Symposium on Computer Science Education (SIGCSE '15). Association for Computing Machinery, New York, NY, USA, 569–574. DOI: https://doi.org/10.1145/2676723.2677289

Radermacher, A., & Walia, G. 2011. Investigating the effective implementation of pair programming: an empirical investigation. In Proceedings of the 42nd ACM Technical Symposium on Computer Science Education (SIGCSE '11). Association for Computing Machinery, New York, NY, USA, 655–660. DOI: https://doi.org/10.1145/1953163.1953346

Radermacher, A., Walia, G., & Rummelt, R. 2012. Improving student learning outcomes with pair programming. In Proceedings of the Ninth Annual International Conference on International Computing Education Research (ICER '12). Association for Computing Machinery, New York, NY, USA, 87–92. DOI: https://doi.org/10.1145/2361276.2361294

Rodríguez, F., Price, K., & Boyer, K. 2017. Exploring the pair programming process: characteristics of effective collaboration. In Proceedings of the 2017 ACM SIGCSE Technical Symposium on Computer Science Education (SIGCSE '17). Association for Computing Machinery, New York, NY, USA, 507–512. DOI: https://doi.org/10.1145/3017680.3017748

Rong, G., Zhang, H., Liu, B., Shan, Q., & Shao, D. 2018. A replicated experiment for evaluating the effectiveness of pairing practice in PSP education. Journal of Systems and Software, 136, 139–152. Science Direct. DOI: https://doi.org/10.1016/j.jss.2017.08.011

Salleh, N., Mendes, E., & Grundy, J. 2010. Empirical studies of pair programming for CS/SE teaching in higher education: A systematic literature review, IEEE Transactions on Software Engineering.Sfetsos, P., Stamelos, I., Angelis, L., et al. 2009. An experimental investigation of personality types impact on pair effectiveness in pair programming. Empirical Software Engineering 14, 187. Springer. https://doi.org/10.1007/s10664-008-9093-5

Salleh, N., Mendes, E., & Grundy, J. 2014. Investigating the effects of personality traits on pair programming in a higher education setting through a family of experiments. Empirical Software Engineering, 19, 714–752. Springer https://doi.org/10.1007/s10664-012-9238-4

Simon, B., & Hanks, B. 2008. First-year students' impressions of pair programming in CS1. Journal on Educational Resources Computing, 7(4), Article 5 (January 2008), 28 pages. DOI: https://doi.org/10.1145/1316450.1316455 ACM

Sison, R. 2009. Investigating the effect of pair programming and software size on software quality and programmer productivity. 2009 16th Asia-Pacific Software Engineering Conference. DOI: 10.1109/apsec.2009.71

Slaten, K., Droujkova, M., Berenson, S., Williams, L., & Layman, L. 2005. Undergraduate student perceptions of pair programming and agile software methodologies: verifying a model of social interaction. Agile Development Conference (ADC'05), Denver, CO, 2005, pp. 323–330. IEEE. DOI: 10.1109/FIE.2005.1611964

Smith, M., Giugliano, A., & DeOrio, A. 2018. Long term effects of pair programming. IEEE Transactions on Education, 61(3), 187–194, Aug. 2018. IEEE DOI: 10.1109/TE.2017.2773024

Stapel, K., KnaussKurt, E., & Becker S. 2010. Towards understanding communication structure in pair programming. In: Sillitti, A., Martin, A., Wang, X., & Whitworth, E. (Eds.). (2010). Agile Processes in Software Engineering and Extreme Programming. Lecture Notes in Business Information Processing. Springer. DOI: 10.1007/978-3-642-13054-0

Stotts D., Williams L., Nagappan N., Baheti P., Jen D., & Jackson A. 2003. Virtual teaming: experiments and experiences with distributed pair programming. In: Maurer F., & Wells D. (Eds.). Extreme Programming and Agile Methods - XP/Agile Universe 2003. XP/Agile Universe 2003. Lecture Notes in Computer Science, vol 2753. Springer, Berlin, Heidelberg. DOI: https://doi.org/10.1007/978-3-540-45122-8_15

Thomas, L., Ratcliffe, M., & Robertson, A. 2003. Code warriors and code-a-phobes: a study in attitude and pair programming. In Proceedings of the 34th SIGCSE Technical Symposium on Computer Science Education (SIGCSE '03). Association for Computing Machinery, New York, NY, USA, 363–367. ACM DOI: https://doi.org/10.1145/611892.612007

Tsompanoudi, D., Satratzemi, M., & Xinogalos, S. 2016. Evaluating the effects of scripted distributed pair programming on student performance and participation. IEEE Transactions on Education, 59(1), 24–31. DOI: 10.1109/te.2015.2419192

Tsompanoudi, D., Satratzemi, M., & Xinogalos, S. 2019. An empirical study on factors related to distributed pair programming. International Journal of Engineering Pedagogy, 9(2), 65–81. International Association of Online Engineering (IAOE), Austria.

Umapathy, K., & Ritzhaupt, A. 2017. A meta-analysis of pair-programming in computer programming courses: implications for educational practice ACM transactions on computing education. ACM Transactions on Computing Education, 17(4), Article 16. https://dl.acm.org/citation.cfm?id=2996201

Urai, T., Umezawa, T., & Osawa, N. 2015. Enhancements to support functions of distributed pair programming based on action analysis. In Proceedings of the 2015 ACM Conference on Innovation and Technology in Computer Science Education (ITiCSE '15). Association for Computing Machinery, New York, NY, 177–182. DOI: https://doi.org/10.1145/2729094.2742616

Van Toll III, T., Lee, R., & Ahlswede, T. 2007. Evaluating the usefulness of pair programming in a classroom setting. 6th IEEE/ACIS International Conference on Computer and Information Science (ICIS 2007), Melbourne, Qld, 2007, pp. 302–308. DOI: 10.1109/ICIS.2007.96

VanDeGrift, T. 2004. Coupling pair programming and writing: learning about students' perceptions and processes. In Proceedings of the 35th SIGCSE Technical Symposium on Computer Science Education (SIGCSE '04). Association for Computing Machinery, New York, NY, 2–6. DOI: https://doi.org/10.1145/971300.971306

Williams, L., Kessler, R., Cunningham, W., & Jeffries, R. 2000. Strengthening the case for pair programming. IEEE Software, 17(4) (July 2000), 19–25. DOI: https://doi.org/10.1109/52.854064

Williams, L., Layman, L., Osborne J., & Katira, N. 2006. Examining the compatibility of student pair programmers. AGILE 2006 (AGILE'06), Minneapolis, MN, 2006, pp. 10 pp.-420. IEEE DOI: 10.1109/AGILE.2006.25

Williams, L., McCrickard, D., Layman, L., & Hussein, K. 2008. Eleven guidelines for implementing pair programming in the classroom. In: Melnik G., & Poppendieck M. (Eds.). In Proceedings of the Agile Conference, IEEE, 445–452.

Williams, L., McDowell, C., Nagappan, N., Fernald, J., & Werner, L. 2003. Building pair programming knowledge through a family of experiments. 2003 International Symposium on Empirical Software Engineering (ISESE 2003). IEEE. pp. 143–152. Rome, Italy. https://doi.org/10.1109/isese.2003.1237973

Wohlin C, Runeson P, Host M, Ohlsson MC, Regnell B, & Wesslen A. 2000. Experimentation in Software Engineering: An Introduction. Kluwer Academic Publishers: Norwell.

Wood, K., Parsons, D., Gasson, J., & Haden, P. 2013. It's never too early: pair programming in CS1. In Proceedings of the Fifteenth Australasian Computing Education Conference – Volume 136 (ACE '13). Australian Computer Society, Inc., AUS, 13–21. ACM

Xinogalos, S., Satratzemi, M., Chatzigeorgiou, A., & Tsompanoudi, D. 2017. Student perceptions on the benefits and shortcomings of distributed pair programming assignments. 2017 IEEE Global Engineering Education Conference (EDUCON), Athens, 2017, pp. 1513–1521. DOI: 10.1109/EDUCON.2017.7943050

Ying, K., Michelle, P., Lydia, G., Ahmed, M. & Crompton, K. 2019. In their own words: gender differences in student perceptions of pair programming. In Proceedings of the 50th ACM Technical Symposium on Computer Science Education (SIGCSE '19). Association for Computing Machinery, New York, NY, 1053–1059. ACM DOI: https://doi.org/10.1145/3287324.3287380

Zacharis, N. 2011. Measuring the effects of virtual pair programming in an introductory programming java course. IEEE Transactions on Education, 54(1), 168–170. DOI: 10.1109/te.2010.2048328

Chapter 14

Programming Multi-Agent Coordination Using *NorJADE* Framework

Toufik Marir[a], Selma Mammeri[a],
and Rohallah Benaboud[a]

[a]*Research Laboratory on Computer Sciences' Complex Systems,
University of Oum El Bouaghi, Oum El Bouaghi, Algeria*

Contents

14.1   Introduction 287
14.2   Related Works 289
14.3   NorJADE Framework..290
14.4   Implementing Coordination Mechanisms Using *NorJADE*....................293
14.5   Case Study 295
14.6   Discussion 299
14.7   Conclusion 301
References 301

14.1 Introduction

Multi-agent software paradigm is a promising software paradigm to develop complex systems. In fact, multi-agent systems are a set of interacting of entities (called agents). This latter is characterized mainly by autonomy, reactivity, pro-activity and social ability (Wooldridge, 2009). Consequently, the agents can produce flexible and intelligent behaviors. Moreover, the interaction between agents is a complex phenomenon that can model reorganization and self-organization situations (Abbas et al., 2015). Both individual and collective characteristics of agents and multi-agent systems provide necessary tools to model complex systems.

Recently, norms become a key concept in multi-agent systems. Norms are rules that guide or constraint agents' behaviors (Cialdini & Trost, 1998). Compared to the functional specifications which represent the ideal behavior of a system, an agent can violate norms by executing prohibited behaviors and undertaking their consequences. Specifying the agents' behaviors using norms, multi-agent systems become more flexible and can behave with more autonomy. An agent can execute prohibited behaviors if the situation requires such decision.

Developing norms is an important issue in agent oriented software engineering. Naturally, proposing new software paradigm implies developing new methods, methodologies and techniques to simplify the development of projects using this paradigm. For example, many methods and methodologies are proposed to accompany developers during the development process of multi-agent systems (Bergenti et al., 2006). Integrating norms with multi-agent systems opened new perspectives to develop more complex systems as it brought out new challenges during the development of such systems. It is important to propose methods to specify, to design, to develop and to test normative multi-agent systems.

Many recent works targeted the problems of the development of normative multi-agent systems. Especially, implementing and programming such systems is a hot topic in this field (Dastani et al., 2009; Grossi et al., 2010; Hübner et al., 2011; Morales et al., 2014; Kafali et al., 2017). However, most proposed works are limited to only a specific kind of norms (called regulative norms). We think that providing frameworks to develop other kinds of norms can really open more perspectives for the application of normative multi-agent systems. Particularly, norms can be used as a base to ensure coordination between agents in order to promote openness and enhance heterogeneity. These attributes (openness and heterogeneity) are very required in recent software as a natural attributes of complex systems. Moreover, using this kind of norms allows updating goal of agents or their coordination protocol without changing the code of the system. Consequently, the developed systems become more flexible.

In this chapter, we propose a new extension of *NorJADE* framework (Marir et al., 2019) to implement coordination mechanisms for multi-agent systems. *NorJADE* (Marir et al., 2019) is an extension framework of *JADE* (Bellifemine et al., 2005) platform to develop normative systems. In the previous version of this framework, we focused on regulative norms and we gave only some basic stones to develop procedural norms. Especially, in this version we focused in implementing collective goals within normative multi-agent systems. We think that norms can be used as a natural solution to represent the coordination mechanism within a society of agents.

The remainder of this chapter is organized as following: Section 14.2 is devoted to present related works. Then, we present briefly *NorJADE* framework (the framework we used as a base to our approach) in Section 14.3. We described the proposed extension of *NorJADE* to implement coordination mechanisms in Section 14.4. In Section 14.5, we applied this approach on a case study. Then, we discussed the advantages of this extension in Section 14.6. Finally, we present conclusion and some perspectives in Section 14.7.

14.2 Related Works

According to Marir et al. (2019), we can distinguish two tracks in the implementing of normative multi-agent systems' field. Firstly, in several works, implementing normative multi-agent systems is made as a secondary goal in the validation contexts of some approaches and mechanisms (Felicíssimo et al., 2008; Morales et al., 2014; Kafali et al., 2017;). In this track, we use general purpose programming languages (like *Java* or *prolog*) in spite of the expression limits of these latter. In fact, these languages do not provide the abstract level to express and to manipulate norms in a natural way.

Secondly, several approaches proposed languages, platforms and frameworks to implement normative multi-agent systems. The main goal of these approaches is to provide an adequate abstraction level to represent and manipulate norms. Knowing that norms are an integrated part of multi-agent organizations, some organizational models proposed solutions to represent and manipulate norms. Hübner et al. (2011) proposed an approach to translate a specification of *Moise* organization model (Hübner et al., 2002) to a normative program that can be interpreted by an organization management infrastructure (OMI). The authors proposed a normative programming language (NPL) as a target language of the translation process. However, the proposed programming language supports only some regulative norms' concepts like regimentation and obligation concept. Moreover, this approach is closely related to *Moise* model which limits its applicability.

Dastani et al. (2009) proposed a simplified language to implement normative multi-agent systems. This language is specified using several normative concepts like monitoring, regimenting and sanction mechanisms. However, all these concepts are related to regulative norms. Moreover, norms are specified as an integrated part of the multi-agent program (endogenous representation). Consequently, updating norms at the run-time will be hard difficult.

Marir et al. (2019) remarked that almost all the existing solutions to implement normative multi-agent systems are complex because they are based on complex theoretical frameworks or they use general purpose languages. Consequently, the authors proposed an extension of *JADE* (Bellifemine et al., 2005) platform to implement normative multi-agent systems (Called *NorJADE*). *JADE* (Bellifemine et al., 2005) is a popular framework to develop multi-agent systems. Using *NorJADE*, we can gain the advantages of *JADE* like simplicity, popularity and maturity (Marir et al., 2019). However, the last version of *NorJADE* focused on regulative norms and procedural and constitutive norms are treated superficially. For example, procedural norms allow only implementing individual goals. If a goal is described as a collective one that requires the coordination of several agents, *NorJADE* framework did not give a solution to implement this situation. Nonetheless, norms can play a fundamental role to ensure coordination between agents. Therefore, in this chapter we present an extension of *NorJADE* framework that allows implementing coordination mechanisms within a multi-agent system.

14.3 *NorJADE* Framework

NorJADE is a new framework for programming normative multi-agent systems (Marir et al., 2019). This framework is based on a popular agent development framework (called *JADE*) using well-known technologies (aspect-oriented programming and ontology).

Figure 14.1 presents the general architecture of *NorJADE* framework. In fact, norms are described in *NorJADE* as ontology. This ontology is manipulated by an extension of *JADE*-based program. This extension is implemented as a library of aspects (coded by *AspectJ* (Kiczales et al., 2001)) to enhance the standard agents by specific capabilities of controlling and manipulating norms.

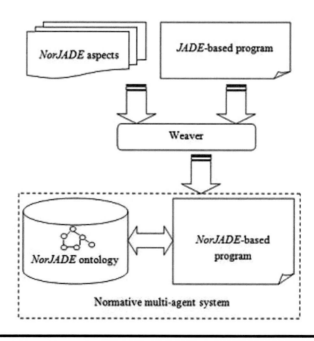

Figure 14.1   The general architecture of *NorJADE* framework. (Marir et al., 2019.)

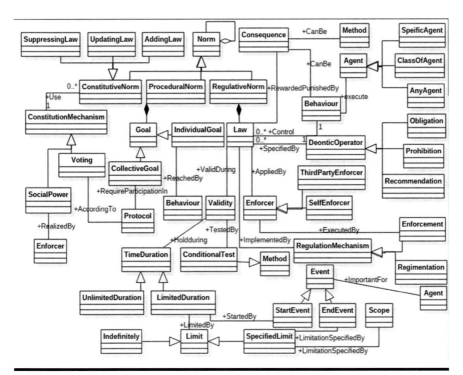

Figure 14.2   The concepts of *NorJADEOntology*. (Marir et al., 2019.)

Most important concepts of normative multi-agent systems' field are described using ontology (called *NorJADEontology*) (Figure 14.2). Hence, three kinds of norms are distinguished (regulative, procedural and constitutive norms). Regulative norms are used to control the agents' behaviors. As we said above, the last version of this framework focused on this kind of norms. Therefore, we can find almost all the concepts allowing the control of agents' behaviors. Mainly, we can mention behavior, deontic operator, enforcer and consequence. Obviously, we use norms to control agents' behaviors using a deontic operator. A deontic operator is used to specify the constraints imposed the specified behavior (like prohibition, obligation or recommendation). According to the normative multi-agent systems' principles, an agent can violate a norm. In this case, a specific agent must control respecting and violating norms. In a violating norms case, a consequence must be executed.

NorJADE can be used also to describe constitutive norms by specifying a constitution mechanism (voting or social power). By constitutive norms, we mean the norms used to create or update norms (Mahmoud et al., 2014). Moreover, the procedural norms are specified by goals. This kind of norms expresses how decisions are made by agents in a normative system (Mahmoud et al., 2014). However, in

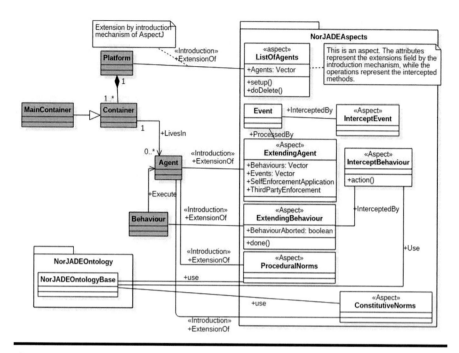

Figure 14.3   The *NorJADEAspects* architecture. (Marir et al., 2019.)

the last version of *NorJADE* framework, only individual goals are described. In this way, an abstraction level is provided allowing agents to update the way to reach their goals (by updating *NorJADEontology*) without updating the code of the multi-agent system.

In order to manipulate norms, *JADE* framework is enhanced by a library of aspects (Figure 14.3). This latter is used to manipulate and reason about norms. Mainly, these aspects specify the monitoring and regulation mechanism. The monitoring mechanism is coded in the *InterceptBehaviour* aspect (Figure 14.4). It allows detecting the execution of behaviors by an agent, and controlling whether or not these behaviors are regulated by a norm. In the case of a regulative behavior, *NorJADE* framework will execute the regulation mechanism. The regulation mechanism consists in aborting a prohibited regulative behavior or executing its consequence (as a reward or a punishment).

Despite that *NorJADE* framework allows programming most concepts of norms, it focuses on regulative ones. Programming procedural norms is not treated deeply. Especially, norms can be used more effectively to ensure coordination between agents. Naturally, the coordination mechanisms can be described as norms to enhance homogeneity between heterogeneous agents. In the next section, we present an extension of *NorJADE* framework that allows programming coordination mechanisms using norms.

```
1   public aspect InterceptBehaviour {
2     public pointcut BehaviourExecuting():execution(void *.action());
3       Object around() : BehaviourExecuting() {
4         //Interrogate the NorJADE ontology about the intercepted behaviour
5         if(RegulatedBehaviour != null) {
6           //Collect information about the law
7           if(RegulationMechanism.equals("Regimentation")){
8             //The intercepted behaviour will be aborted
9                   }
10        else{
11            // The intercepted behaviour will continue its execution
12            switch (DeonticOperator){
13              case "Prohibition":{
14              // The punishment will be executed
15              }
16            case "Recommendation":{
17              // The reward will be executed
18              }
19              case "Obligation":{
20            //The intercepted behaviour will be saved for later processing
21              }
22            }
23          }
24        else{
25          //The behaviour will continue its execution
26          }
27      }
28    }
```

Figure 14.4   A portion of the code of *InterceptBehaviour* aspect. (Marir et al., 2019.)

14.4 Implementing Coordination Mechanisms Using *NorJADE*

In order to allow implementing coordination mechanism within a society of agents, we developed an extension of *NorJADE* framework. Obviously, we followed the same principles of last version of *NorJADE* framework: describing norms as ontology and manipulating them thanks to the aspect-oriented programming. Ontology allows a representation of conceptualization in an unambiguous language. Guarino (1992) defined an ontology as "*a representation of an engineering artifact, consisting of a specific vocabulary used to describe a certain reality, accompanied by a set of implicit assumptions about the meaning of the words in that vocabulary*". It provides an explicit, formal and exogenously representation of norms (Marir et al., 2019); besides, it is sharable, which is very suitable for representing coordination mechanism between agents. Whereas aspect-oriented programming is relatively a new programming paradigm (Kiczales et al., 1997). It allows separating crosscutting concerns to core ones in order to improve the reusability of software. The crosscutting concerns can be developed independently as aspects thanks to this software paradigm. Using the aspect-oriented programming, we can extend *JADE* framework by the required mechanism to manipulate norms without updating the core of the initial framework. Consequently, developers familiar with *JADE* can use naturally *NorJADE* (Marir et al., 2019).

NorJADE ontology is extended by several concepts related to coordination mechanisms. Coordination mechanisms are applied when we have a collective goal. Hence, this concept is the first stone in the ontology (as is presented in Figure 14.5). Reaching this goal requires participating in protocols. A standard protocol can be described by its name and the performative of its first message. In fact, protocols in multi-agent systems' field are not only described in the specialized literature, but they are also standardized by FIPA organization (FIPA, 2002). Usually, a protocol will be initiated by an initiator and require participation of one or several responders. An initiator is an agent that will execute the initiator behavior. On the other side, responder is an entity that will participate in the protocol. This latter can be described as an agent or as a service. In fact, sometimes the initiator knows the agents participating in the coordination situation. However, in many other cases (especially, in the case of open systems), the initiator does not know the responder agents, but it knows only the required services from these agents. In this case, the responder will be described as a service. Obviously, a responder is in charge of lunching the responder behavior.

Describing the coordination mechanisms as ontology instead of integrating them in the code of multi-agent systems provides several advantages. First, agents can reason about the coordination mechanisms before its commitment because this description is formal and explicit. Secondly, this description is shared within a society of agents. Consequently, it is not necessary that agents know beforehand the used protocol. Particularly, it is hard difficult to specify the used protocol in the case of open systems where agents can joint and leave the society of the agent during its runtime. Finally, using ontology, we can update the used protocol to ensure the coordination mechanism even during runtime without updating the code of the multi-agent system.

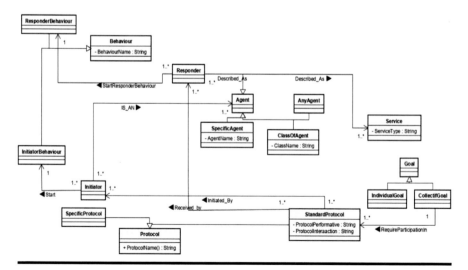

Figure 14.5   An extension of *NorJADEontology* to describe coordination mechanisms.

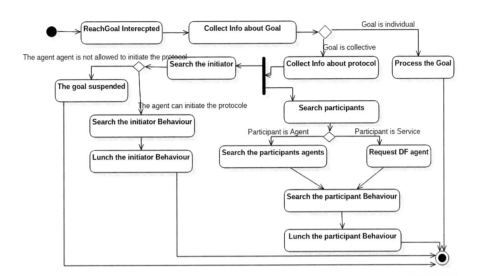

Figure 14.6   The activity diagram of the extended aspect.

In addition to the extension of the ontology described above, we also extended the library of aspects by an aspect that describes the operational aspect of reaching a collective goal. Figure 14.6 presents the activity diagram of this aspect. In fact, we used the introduction mechanism of *AspectJ*, to enhance the standard *JADE* agent with the method **ReachGoal**. When this method is executed, the aspect collects the information about this goal. In the case of an individual goal, the associated behavior will be executed. On the other hand, if the goal is collective one, the aspect collects the information about the required protocol to reach this goal using *NorJADE* ontology. The result of this process is information about the initiator and information about the participants. Obviously, if the agent that executed the **ReachGoal** method is not allowed to lunch the protocol, this latter will be suspended. Otherwise, the agent lunches the initiator behavior. At the same time, the aspect searches the specified participants. As we said in the ontology, a participant can be described as an agent or as a service. If it is described as an agent, then all the participant agents will be found and will execute their response behavior. Else, a request will be sent to the DF agent to identify the agents which can provide the specified service. Afterward, these agents will execute their response behavior. Figure 14.7 gives a portion of the code of this aspect.

14.5 Case Study

In order to validate our approach and to clarify their advantages, we applied it on a case study with several execution scenarios. In fact, we applied our approach on the conference management system. More precisely, we implemented different

```
 2⊕ import java.lang.reflect.Constructor;▯
 19  public aspect CollectiveGoal {
 20⊝     static //pointcut test() :   call( static String *.getprotocol(..));
 21  String tabAgent[]=new String[200];
 22      public static String GoalClass;
 23⊝     public  void Agent.ReachGoal(String goal){
 24          //ACLMessage msg=new ACLMessage(3);
 25          GoalClass=requests.MyRequests.getGoalClass(goal);|
 26          if (GoalClass.equals("CollectiveGoal")) {
 27              String Protocol=requests.MyRequests.getProtocol(goal);
 28              String classProtocol=requests.MyRequests.getProtocolClass(Protocol);
 29              if (classProtocol.equals("StandardProtocol")) {
 30                  String nameProtocol=requests.MyRequests.getProtocolName(Protocol);
 31                  String performative=requests.MyRequests.getProtocolPerformative(Protocol);
 32                  String initiatorAgent=requests.MyRequests.getInitiatorAgent(Protocol);
 33                  String agent=requests.MyRequests.getAgent(initiatorAgent);
 34                  String ClassAgent=requests.MyRequests.getClassAgent(agent);
```

Figure 14.7 Portion of the code of the aspect *CollectiveGoal*.

scenarios to dispatch papers to reviewers. It is important to note that the conference management system is the typical example of applying normative multi-agent systems because it provides several kinds of norms (Luck & Padgham, 2008; Marir et al., 2019). In our case, we choose to apply our approach using several scenarios: the case of participant described as agent, the case of participant described as service and in the case of updating the protocol.

In the first case, we specified the coordination mechanism as the protocol *ContractNet*. This protocol is usually used in the case of cooperation situations between agents to identify the best propose for achieving a task. As it is presented in Figure 14.8, the initiator sends a *CFP* to several participants. A participant can refuse doing this task or propose a price to do it. After receiving all the responses of participants, the initiator evaluates them and it sends an accept-proposal to the best proposal and reject-proposal to other agents. The participant agent informs the initiator by the task execution result. In our case, the initiator is the chairman of the conference and the participants are the reviewers. Each reviewer proposes its level of qualification to review a paper as a response to the *CFP* of the chairman. As it is presented in the Figure 14.9, the chairman sent its *CFP* to four reviewers. One reviewer refused participating, and 03 ones gave their proposals. The chairman chose the best proposal (the proposal of the agent 3), which then failed in reaching the goal.

Assuming now that the conference is a multidisciplinary one. In this case, the chairman has to identify the possible reviewers according to their areas of interest. Consequently, the participant is described in the ontology as a service. Using our extension of *NorJADE* framework, we will identify the possible reviewers according to the specified service using DF agent of *JADE*. Then, we will apply the *ContractNet* protocol as is presented in the first scenario. Figure 14.10 presents the result of the execution of this second scenario. In this scenario, the reviewer 3 refused to participate and the chairman chose the reviewer 0 to accomplish the task. This latter accomplished the task successfully.

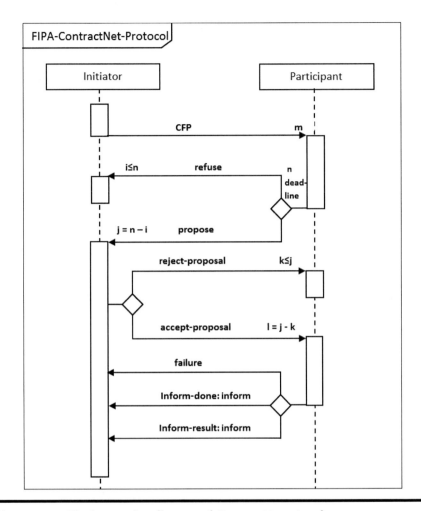

Figure 14.8 **The interaction diagram of** *ContractNet* **rotocol.**

In the last scenario, we validate the possibility of changing the coordination mechanism without updating the code of the multi-agent system. In fact, we only change the protocol's name in the ontology from *ContractNet* to *Request* protocol. As is presented in Figure 14.11, in this protocol the initiator requests a service from the participant. This latter can refuse or accept accomplishing the task. If he accepted, he will inform the initiator by the result of the task execution.

In the case of our case study, the chairman does not consult reviewers to identify the best ones to review a paper, but he has to select a reviewer and assign him a paper for reviewing. Obviously, the reviewer can accept or refuse this task. If he accepted, he must inform the chairman by the result of the reviewing process. Figure 14.12 presents the result of the execution of this scenario. In fact, two

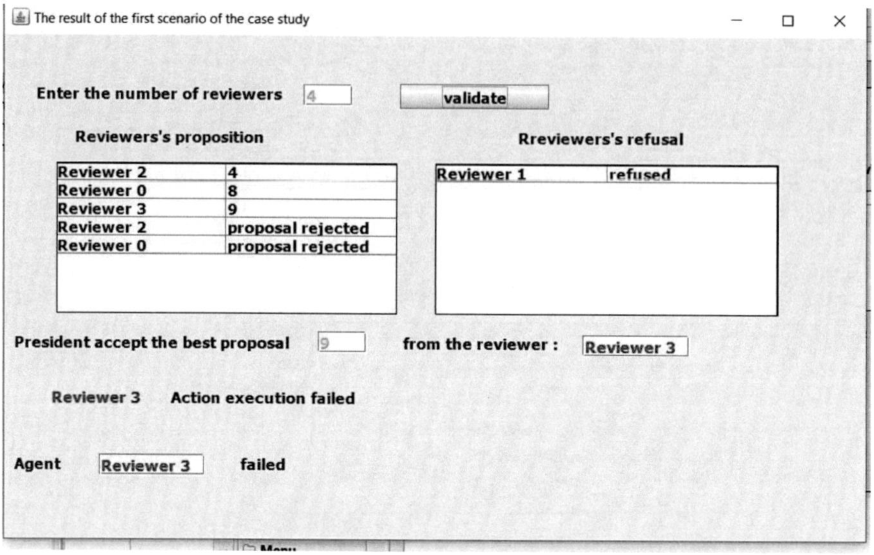

Figure 14.9　The result of the first scenario of the case study.

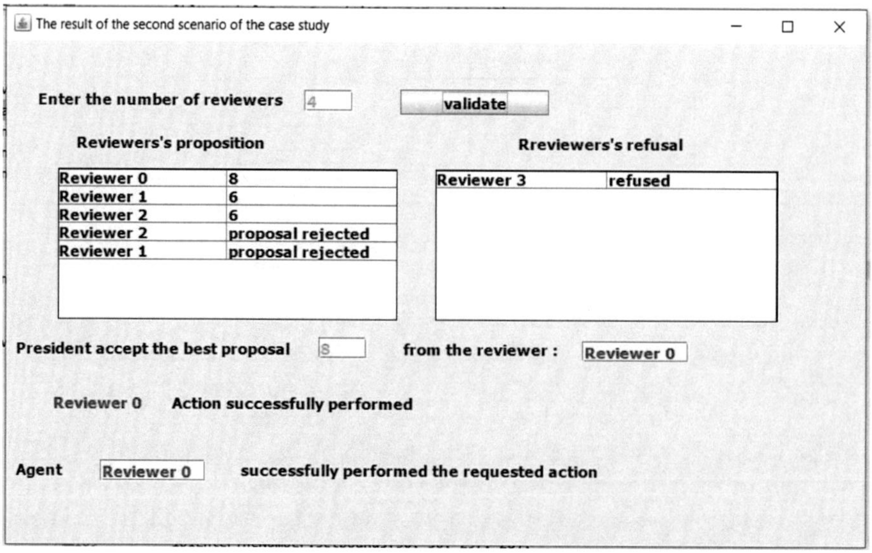

Figure 14.10　The result of the second scenario of the case study.

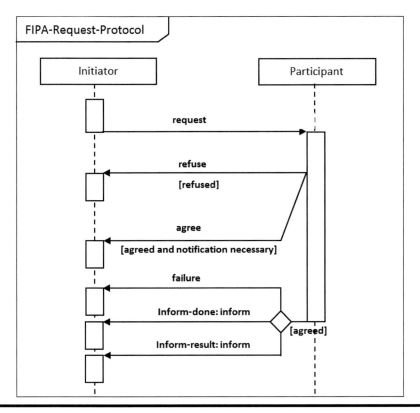

Figure 14.11 **The interaction diagram of the *Request Protocol*.**

reviewers are refused to participate in the coordination protocol (Reviewer 2 and Reviewer 3) and two others are accepted. Then, only the reviewer 1 informed the chairman that he has successfully accomplished his task.

14.6 Discussion

Developing an extension of *NorJADE* to implement coordination mechanisms for multi-agent systems provide several advantages. Firstly, as an extension of *NorJADE*, the developed framework provides the same advantages of *NorJADE* like simplicity, reusability, genericity and extensibility (Marir et al., 2019). In fact, *NorJADE* is based on a simple and popular multi-agent framework (called *JADE*) also it is based on well-known technologies (like aspect-oriented programming and ontologies). Moreover, *JADE* framework is a generic one because it is not based on a specific agent model. In addition, *NorJADE* is extensible thanks to aspect-oriented programming that allows adding more aspects in a naturally, and it provides reusable components to develop normative multi-agent systems.

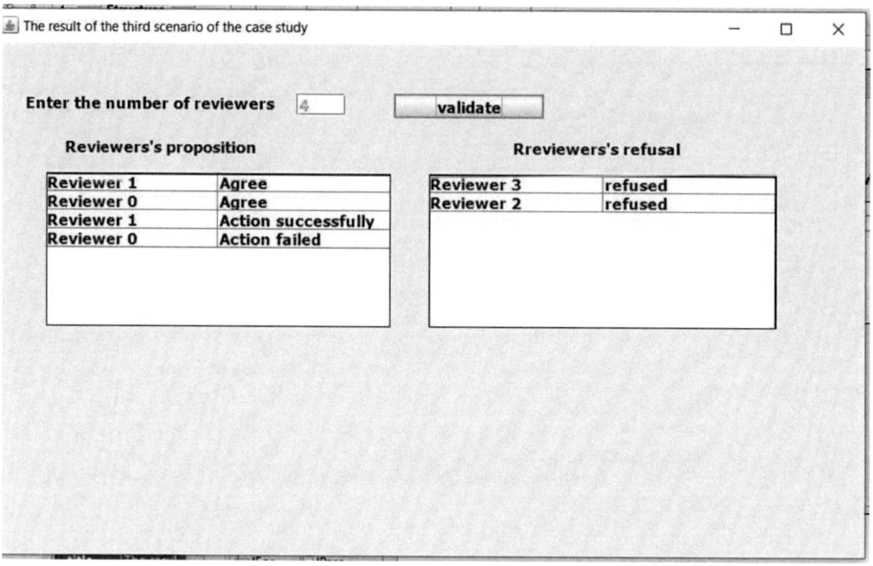

Figure 14.12 The result of the third scenario of the case study.

Secondly, this extension of *NorJADE* allows programming coordination mechanisms as it is known in real life (a metaphor of real life). In fact, coordination in human societies is based generally on rules (called laws, norms, ethics, standards or traditions according to the contexts). Consequently, we think that is more beneficial to represent the coordination as it is described in real life instead of transforming it to other models.

In addition, describing the coordination mechanisms as an external, explicit and formal part (as a norm) instead of described it in the code of the multi-agent system (as a protocol) allows developing open systems. In this kind of systems, agents can join and leave the system during runtime. Consequently, agents did not know beforehand the applied coordination mechanism to reach a collective goal. Contrariwise, they should adapt their behaviors according to the coordination mechanisms of each system. Therefore, the agents will consult and reason about these coordination mechanisms which are described as norms.

Furthermore, this extension of *NorJADE* enhanced the maintainability of multi-agent systems. As it is presented in third scenario, we can update only the ontology in order to update the coordination mechanisms. Usually, the coordination mechanisms are described as protocols and coded within the multi-agent code. Consequently, updating this code that includes individual and collective tasks of the agents is a difficult. However, updating the ontology is easier.

Finally, this extension of *NorJADE* enhanced the flexibility of multi-agent systems. In fact, it is possible to reason and update the ontology by agents during runtime. Consequently, agents can adapt their coordination mechanisms according to the current situation.

14.7 Conclusion

Recently, norms play an important role in multi-agent systems. Indeed, several recent works propose approaches to include norms in multi-agent systems like models of normative multi-agent systems, processes for creating and analyzing conflicts among norms, approach for describing and reasoning about norms. In this context, we work on the engineering of normative multi-agent systems. Our goal is providing artifacts and techniques that simplify the development of normative multi-agent systems. *NorJADE* is one of these artifacts. It is an extension of *JADE* to develop normative multi-agent systems. In this work, we extended *NorJADE* to support programming coordination mechanisms for multi-agent systems. Hence, an ontology of coordination mechanisms is developed. In addition, an aspect is implemented to represent the operational aspect of the coordination mechanism (manipulation of the ontology). This approach is validated using the conference management system as a case study. We think that our approach provides several advantages like flexibility and maintainability. Also, it is based on real life metaphor and it is more suitable for open systems.

As a perspective, we think that it is important to propose a language to describe protocols using the ontology. This version of *NorJADE* is limited to standards protocols. Enhanced this ontology by describing protocols allows updating protocols during runtime which enhances the flexibility. Also, we think that it is important to develop constitutive norms and related them to the coordination mechanisms.

References

Abbas, H. A., Shaheen, S. I., & Amin, M. H. (2015). Organization of multi-agent systems: an overview. International Journal of Intelligent Information Systems, 4(3), 46–57.

Bellifemine, F., Bergenti, F., Caire, G., & Poggi, A. (2005). JADE—a java agent development framework. In *Multi-Agent Programming* (pp. 125–147). Boston, MA: Springer.

Bergenti, F., Gleizes, M. P., & Zambonelli, F. (Eds.). (2006). *Methodologies and Software Engineering for Agent Systems: The Agent-oriented Software Engineering Handbook* (Vol. 11). Springer Science & Business Media, Kluwer academic publishers, Boston.

Cialdini, R. B., & Trost, M. R. (1998). Social influence: Social norms, conformity and compliance. In D. T. Gilbert, S. T. Fiske, & G. Lindzey (Ed.), *Handbook of Social Psychology* 2 (pp. 151–192). NewYork: Oxford University Press.

Dastani, M., Tinnemeier, N. A., & Meyer, J. J. C. (2009). A programming language for normative multi-agent systems. In *Handbook of Research on Multi-agent Systems: Semantics and Dynamics of Organizational Models* (pp. 397–417). IGI Global.

Felicíssimo, C. H., Briot, J. P., Chopinaud, C., & de Lucena, C. J. P. (2008, July). DynaCROM: an approach to implement regulative norms in normative multiagent systems. In 3rd International Workshop on Normative Multiagent Systems (NorMAS'08), Jul 2008, Luxembourg, Luxembourg. (pp. 80–94).

FIPA (2002) http://www.fipa.org/repository/ips.html

Grossi, D., Gabbay, D., & Van Der Torre, L. (2010). The norm implementation problem in normative multi-agent systems. In *Specification and Verification of Multi-agent Systems* (pp. 195–224). Boston, MA: Springer.

Guarino, N. (1992). Concepts, attributes and arbitrary relations: some linguistic and ontological criteria for structuring knowledge bases. Data & Knowledge Engineering, 8(3), 249–261.

Hübner, J. F., Boissier, O., & Bordini, R. H. (2011). A normative programming language for multi-agent organisations. Annals of Mathematics and Artificial Intelligence, 62(1-2), 27–53.

Hübner, J. F., Sichman, J. S., & Boissier, O. (2002, November). A model for the structural, functional, and deontic specification of organizations in multiagent systems. In *Brazilian Symposium on Artificial Intelligence* (pp. 118–128). Berlin, Heidelberg: Springer.

Kafali, O., Ajmeri, N., & Singh, M. P. (2017, February). Kont: Computing tradeoffs in normative multiagent systems. In *Thirty-First AAAI Conference on Artificial Intelligence* (AAAI'2017), AI Access Foundation, San Francisco, United States, (pp. 3006–3012).

Kiczales, G., Hilsdale, E., Hugunin, J., Kersten, M., Palm, J., & Griswold, W. G. (2001, June). An overview of AspectJ. In *European Conference on Object-Oriented Programming* (pp. 327–354). Berlin, Heidelberg: Springer.

Kiczales, G., Lamping, J., Mendhekar, A., Maeda, C., Lopes, C., Loingtier, J. M., & Irwin, J. (1997, June). Aspect-oriented programming. In *European Conference on Object-oriented Programming* (pp. 220–242). Berlin, Heidelberg: Springer.

Luck, M., & Padgham, L. (Eds.). (2008). Agent-oriented software engineering VIII: 8th International Workshop. In *AOSE 2007*. Honolulu, HI: Springer. May 14, 2007, Revised Selected Papers (Vol. 4951).

Mahmoud, M. A., Ahmad, M. S., MohdYusoff, M. Z., & Mustapha, A. (2014). A review of norms and normative multiagent systems. The Scientific World Journal, 2014.

Marir, T., Silem, A. E. H., Mokhati, F., Gherbi, A., & Bali, A. (2019). NorJADE: an open source JADE-based framework for programming normative multi-agent systems. International Journal of Open Source Software and Processes (IJOSSP), 10(2), 1–20.

Morales, J., Mendizabal, I., Lopez-Sanchez, M., Wooldridge, M., & Vasconcelos, W. (2014, May). Normlab: A framework to support research on norm synthesis. In *Proceedings of the 2014 International Conference on Autonomous Agents and Multi-agent Systems* (pp. 1697–1698). International Foundation for Autonomous Agents and Multiagent Systems.

Wooldridge, M. (2009). *An Introduction to Multiagent Systems*. John Wiley & Sons Ltd, United Kingdom.

Index

Note: Locators in *italics* represent figures and **bold** indicate tables in the text.

A

Aarne, O., 272
Accuracy, 64, 66–68
ACM digital library, 259
Adlemo, A., 61, 62
Akinator (game), 182
AliCe-ViLlagE platform, 275
Ali, M. S., 4
Al-Jarrah, A., 275
Alternative hypothesis, 77
Apollo Syndrome, 97
Application program interface (API) patterns, 203
Artificial Neural Network (ANN) technique, *225*, 225–226
AspectJ
    logging implementation, 4
    parallel measurements, 29–30
        Java 1.8 and AspectJ 1.8, 9, *11–12*, 12
        Java 1.9 and AspectJ 1.9, 12, *14*, 16, *16*
    versions, 3, 4–5
Aspect-oriented programming, 3–4; *see also* Execution tracing
Association for Computing Machinery (ACM), 119
Attitudes Towards Mathematics Inventory (ATMI), 264
Automotive software testing
    benefits, test automation, 136
    future work, 154
    limitations, 136
    quality issues, 143–144, **145**
    results
        applied research strategies, 142, *142*

type of publication over the years, 142, *142*
    systematic literature review
        data analysis, 141, **141**
        data extraction, 140–141
        impurity removal, 140
        inclusion and exclusion criteria, 140
        initial search, 139–140
        merge and duplicate removal, 140
        paper selection process, 138–139, *139*
        phases, 137, *137*
        search-and-selection process, 138–140, *139*
    targeted problems, 146, **146**
    test case prioritization, 134–135, *135*
        methods used, 138, 147, 152
        optimization, 138, 152–153
        publication trends, 138, 147
        quality evaluation, 138, 152
    testing methods, common techniques, 143, *143*, **144**
    validation of selected studies, **146**, 146–147, **148–151**
    validity threats, 153
Autonomy, 65, 69

B

Bag of words (BOW) models, 160, 175
Baheti, P., 274
Base model, 78
Bellini, E., 262, 271
Bettin, B., 277
Between–subjects factor, 121
Bipp, T., 265
Black-box testing, 152
Blocker bugs, 163
BM25 (REP), 175
Bowman, N., 271

Boyer, K., 264
Bug life cycle, *162*, 162–163
Bug report, *162*, 162–163
Bug tracking systems
    CNN architecture *vs.* traditional machine
        learning models
        Doc2vec features, **173**
        micro-average F-measure, *173*
        three state-of-the-art models, **174**
    deep learning architectures
        CNN architecture, 169
        hybrid CNN-LSTM/GRU, 170
        LSTM/GRU, 169
    evaluating effectiveness
        datasets, 170
        metrics, 170–171
    results
        deep learning architectures, performance
          of, **171**, 171–172
        machine learning techniques, 172
        micro-average F-measure, 172, *172*
    severity identification, 160–161
        embedding, 168
        preprocessing, 167–168
    usage, 160
Bugzilla, 160, 163

C

Capability Maturity Model (CMM), 92
Carver, J. C., 266
Cataldo, M., 84
Celepkolu, M., 264
Change impact, 70
Chan, W. K., 204
Charters, S., 275
Chaturvedi, K., 175
Chen, K., 254, 264, 270, 272, 276
Choi, K., 270, 272
Clarity, 63, 69
Cohen's kappa metric value, 257
Collaboration independent variable, 79
Collaborative development environment
    (CDE)
    DDN, 78, *78*
    definition, 76
    hypotheses validation, 82
    modeling paradigm, choice of, 80
    model variables
        base model, 78
        control variable, 79–80
        dependent variable, 79

        independent variable, 79
        refined model, 78
    novel perspective, 83
    related literature, 84–85
    results, 80–82, **81**
        utility, 82–83
    validity threats, 83–84
CollectiveGoal method, *296*
Collocated PP (CPP), 274
Colyer, A., 29
Comella-Dorda, S., 61
Common attributes collection, 186, *186*
Completeness, 64, 67–68
Conciseness, 65, 69
Conclusion validity, 275
Connection control variable, 79
Consistency, 64, 68–69
Constitutive norms, 291
Constructive Cost Model (COCOMO), 94
Construct validity, 84, 275
Content validity, 72
Continuous Bag of Words (CBOW), 163
ContractNet protocol, 296–297, *297*, 299
    first scenario, *298*
    second scenario, *298*
    third scenario, *300*
Controlling, 93
Convolutional neural network (CNN), 161,
        164–165, *165*, 172, 175
Correctness, 63, 68
Costa, J. M., 85
Coverability, 64, 68–69
Critical bugs, 163
Cyclomatic complexity (CC), 112

D

DAR algorithm, 203
Da Silva Estacio, B., 254
Dastani, M., 289
Data extraction validity, 72
Decision trees, 182
Deep learning architectures; *see also individual*
        architectures
    CNN architecture, 169
    hybrid CNN-LSTM/GRU, 170
    LSTM/GRU, 169
Defect density (Defects/KLOC), 263
Defect discussion network (DDN), 78, *78*
Dependent research variables, 7–8
Dependent variable, 79
Dissimilarity, 69

Distance control variable, 80
Distributed pair programming (DPP), 254, **273**, 273–274
Diversity, 69
Django, 208, 227
DrupalFeatureFaults.csv, 111
DrupalFeaturesData.csv, 111
Drupal framework
 dataset, 109, **110**, 113
 data value, 111
 description, 111
 feature model, 108
 history of faults, 109, 112–113
 modules, 108, 111–112
 non-functional data, 108–109, 112
Drupal Git repository, 108, 111
DRUPALv4FAMA.xml, 111
DRUPALv4SXFM.xml, 111

E

Eclipse, 170
Efficiency, 64–65, 68
Electronic Control Units (ECU), 152
El-laithym, A., 175
Elsen, E., 203
EoDP, 121, 122
ERASMUS, 121, 122
Erdei, R., 273
Execution tracing, 2–3
 articulated guidelines, 30
 data distributions, each test suite, 8, 33
 parallel measurements, 29–30
  Java 1.8 and Aspect J 1.8, 9, *11–12*, 12, **34**, *35*, **36**, *37*, **38**, *39*, **46**, *47*, **48**, *49*, **50**, *51*
  Java 1.9 and AspectJ 1.9, 12, *14*, 16, *16*, **40**, *41*, **42**, *43*, **44**, *45*, **52**, *53*, **54**, *55*, **56**, *57*
 reliability, 29
 research variables, 7–8
 runtimes comparison with switched on/off states, 16, 18, *18*
  aspect-oriented test application, 22, *24–25*, 25
  both test applications, 27–28, *28*
  conventional test application, 18–19, *19*, *21–22*
 test applications
  sequence diagram, 5–6, *6*
  use case diagram, 5, *5*
 validity, 29

External validity, 84
Extraneous research variables, 7

F

False negative (FN), 227, *227*
False-positive (FP), 227, *227*
FAMIX-based software ontology model, 203
Faults, Drupal, 109, 112–113
Feature interface graph (FIG), 204
Feature models, 108
Fischer, S., 109
Five Factor Model (FFM), 270
Fleming, S. D., 262, 272
F-measure metrics, 170–171
Focus control variable, 80
Fulldataforancova.sav, 121
Fulldata.sav, 121

G

Gated recurrent units (GRU), 161, 167, 172
Gensim library, 168
GitHub repository, 128, 205, 206, *206*, 235, 239
GNOME Project, 77
Goal-Question-Metric (GQM) approach, 61
Gold-Veerkamp, C., 268, 271
Google BigQuery, 206

H

Hamdy, A., 175
Hanks, B., 254, 265, 268, 274
Hard skills, 94–95
Hardware-In-the-Loop (HIL) simulations, 152
Hattori, L., 203
Herbsleb, J. D., 84
Heterogeneous pairs, 270
Hidden Markov models, 129
Hierons, R. M., 109
Highly-Configurable Systems (HCSs)
 examples, 108
 testing, Drupal framework
  dataset, 109, **110**, 113
  data value, 111
  description, 111
  feature model, 108
  history of faults, 109, 112–113
  modules, 108, 111–112
  non-functional data, 108–109, 112

Hilsdale, E., 3
Hinds, P., 85
Homogenous pairs, 271
Hübner, J. F., 289
Hugunin, J., 3
Hussain, S. M., 61, 62

I

IBM Watson, 180
IEEE Xplore, 138, 139, 259
Image collection, 186, *187*
Importance control variable, 80
Inclusion and exclusion criteria, **256**,
    256–257
Independent research variables, 7
Independent variable, 79
Information retrieval (IR) model, 204
Integration testing, 152
InterceptBehaviour aspect, 292, *293*
InterestSpan control variable, 80
Internal validity, 84, 275
Inter-parameter DEpendencies in web Apis
        (IDEA) dataset
    API repository, 105
    data description, 104
    data value, 104
    experimental design, 105
    service-oriented computing, 104
    specification, **103**
Inter-parameter dependency
    constraint, 102
    IDEA dataset (*see* Inter-parameter
        DEpendencies in web Apis (IDEA)
        dataset)
    validation, web APIs, 102
Isong, B., 262
ISPARQL, 203

J

JADE platform, 288, 289, 299, 301
Jarratt, L., 272
Jones, D. L., 262, 272
Jovanovikj, I., 61

K

Kaurkuttal, S., 264, 272, 274, 278
Kiefer, C., 203
Kitchenham, B., 275
K-NN, 175

Knowledge graph, 181–182
Kwan, I., 84

L

Lamkanfi, A., 174
LathaMaheswari, T., 61, 62
Layman, L., 270
Legacy system
    brittle, 62
    clean interfaces, lack of, 63
    definition, 60
    evolution, 60
    expensive maintenance, 63
    finding expertise, 63
    incomplete/outdated documentation, 62–63
    inflexible, changes, 62
    maintenance techniques, 60, 61
    modernization techniques, 60, 61
    obsolete hardware, 62
    organization's backbone, 62
    outdated technologies, 63
    quality testing, 60, 61 (*see also* Quality
        testing)
    replacement techniques, 60
    survey participants
        age of the applications worked on, 66, **67**
        domains of software, 66, **67**
        roles in the projects, 66, **66**
        work experience, 65–66, **66**
    validity threats, 72
Likert scale, 65
Lines of code (LoC), 112
Long short term memory (LSTM), 161, 166,
    172

M

Madeyski, L., 263
Maintenance techniques, legacy systems, 60,
    61, 64, 68
Major bugs, 163
Mann-Whitney test p-values, 9, **10**, **13**, **15**, **17**,
    20, **23**, **26**, **28**
Marcus, A., 174
Marir, T., 289
McDowell, C., 262, 263, 271
McGrath, C., 85
Melis, M., 262
Mendes, E., 262
Mental model consistency (MMC), 270
Menzies, T., 174

Meta-analysis, 117; *see also* MetaMind; Pooled
data meta-analysis
MetaMind dataset
  data collection and transformation,
      117, *117*
  data description, **120**, 120–121
  data value, 119
  specification, **118–119**
Mikolov, T., 163
Mind#1 experiment, 116, 119
Mind#2 experiment, 116
Mind#3 experiment, 116–117
Mindfulness, 116; *see also* MetaMind dataset
  experimental design, 121–122
  software engineering activities, benefits
      of, 122
Mining and API Usage Patterns (MAPO),
      203
Mining software repository (MSR), 200–201
Mining unstructured data
  collection and analysis, 206–207, *207*
  computational model, 225–227, *226*
  design specification
      activity diagram, 221–222, *222*
      architectural description, 221, *221*
      class diagram, 223, *223*
      sequence diagram, 222–223, *223*
  frequently used programming languages,
      232, *235*, **244**
  GitHub repository, 205, *206*
  hybridizing machine learning and NLP
      techniques, 245
  model description, *224*, 224–225, *225*
  modeled classes and methods, 204
  NLP technique, 204–205 (*see also* Natural
      language form (NLP) technique)
  results
      author's log commits, 232, **233**
      execution time testing, *237*
      main components, *231*, 231–232, **236**
      most active pairs, **237**
      quarterly Git commits changes, 232,
          232, **240–241**
      Ruby file on quarterly basis, 232, *234*,
          242–243
      structural flow of changes per week, 232,
          233, **236**
      structured addition from author, 232,
          234, **238–239**
  SAMLRS
      analysis of TP, FP, FN and TN, **227**
      ANN-based, *226*

code executed, determining execution
    time, *231*
  determining, execution time, *230*
  home page interface, *229*
  performance evaluation, 229
  result, repository link, *230*
  Top-N recommendation algorithm, 227,
      228
  user page interface, *229*
  software repositories, 202–203
  system activity diagram, 221–222, *222*
  tools and techniques, 204
Minor bugs, 163
Mixed factorial design, 121
Mixed-gender pairing, 272
Modernization techniques, legacy systems,
    60, 61
Moise organization model, 289
Motivating, 93
Mozilla, 170
Müller, M., 265
Multi-agent software paradigm, 287–288
Multi-agent systems; *see also* NorJADE
        framework
  normative systems, 288
  norms, 288, 291
  proposed languages, 289
Myers–Briggs personality scale, 270
Myers–Briggs Type Indicator (MBTI), 97

N

Naive Bayes (NB), 174
Naive Bayes Multinomial (NBM), 174–175
Natural language form (NLP) technique,
    204–206
  definition, 180, 181
  hybridizing with machine learning,
      245
  parameters, extracting data, **208**
  Predict who (*see* Predict who)
  processed data, **209–220**
  source code, extracting data, *207*,
      207–208
  splitting-function, 208
  stop-words algorithms, 208
  tokenization, 208
Nawahdah, M., 264
Neural network models, 160–161
Nirmala, D., 61, 62
NLP technique *see* Natural language form
    (NLP) technique

Non-functional Drupal data, 108–109, 112
NorJADE framework, 288, 289–290
    activity diagram, extended aspect, 295, *295*
    advantages, 299–300
    ContractNet protocol, 296–297, *297*, 299
        first scenario, *298*
        second scenario, *298*
        third scenario, *300*
    coordination mechanisms, 293–295, *294*,
        295
    general architecture, 290, *290*
    manipulate norms, 292, *292*
    NorJADEAspects architecture, *292*
    ontology, 291, *291*, 294
    Request protocol, 297
        interaction diagram, *299*
Normal bugs, 163
Normative programming language (NPL), 289
Nosek, J., 252
Null hypothesis, 77

O

OpenAPI Specification (OAS), 102, 104
Organization management infrastructure
    (OMI), 289

P

Pair Programming (PP)
    academic performance, 254
    benefits, 252
    concerns
        instructors' role, 269
        learning style, 268
        pair dynamics, 268–269
        personality traits, 268
        technical issues:, 269
    data extraction, 257
    definition, 252
    educational context, 252–254
    factors determining effectiveness, 269–273,
        271
    geographically dispersed partners, **273**,
        273–275
    issues, 275–277
    measures of interest
        effectiveness, 264–266, **265**, **266**
        learning/understanding, 262
        negative feedback, 264
        productivity, 262–263
        quality of work, 263

    student achievements, 265, **265**
    student retention, 263, 266, **266**
    students' perceptions, 263–264
    students' performance, 262
    reduces workload, 273
    results
        list of articles references, 259, **260–261**
        stage wise article selection, 258–259, **259**
    search strategy, 255–257, **256**
    selection criteria, 255–257, **256**
    strategic aspects, 255
    students' perceptions, 266–269, **267**
    thematic analysis, 257–258, **258**
    validity threats, 275
Paper selection process, 138–139, *139*
Paragraph Vector - Distributed Bag of Words
    (PV-DBOW), 164
Paragraph Vector distributed memory
    (PV-DM), 164
Planning, 93
Pontelli, E., 275
Pooled data meta-analysis, 117
Popularity score collection, 187–188, *188*
Population validity
    application age, 72
    technology diversity, 72
POS tagging, 186–187, *187*
Precision metrics, 170–171
Preconditions, 69
Predict who
    adaptiveness, 194
    future work, 194–195
    implementation, 184
        common attributes collection, 186, *186*
        image collection, 186, *187*
        popularity score collection, 187–188,
            188
        POS tagging, 186–187, *187*
        pseudocode, **184–185**
        tokenization, 186–187, *187*
    proposed system, 182–184, *183*
    result criteria, 193–194
    searching process, 193–194
    working, 188–193, *188–193*
Prikladnicki, R., 254
Procedural norms, 291
Process performance indicators (PPIs)
    aligning natural language descriptions, 129
    barrier, 126
    dataset, **126–127**
    data value, 128
    definition, 125–126, 128

description, 128
modeling of PPIs, 129
obtaining the PPI dataset, 129
performance dimensions, 126
public datasets, 130
Punter, T., 60, 65
PyDriller, 206

Q

Quality issues (QI), 143–144, **145**
Quality testing, 60
    auto generation, test cases, 61–62
    data inconsistency, 71
    data loss and recovery, 71
    qualities, test specifications
        accuracy, 64, 67–68
        autonomy, 65, 69
        change impact, 70
        clarity, 63, 69
        completeness, 64, 67–68
        conciseness, 65, 69
        consistency, 64, 68–69
        correctness, 63, 68
        coverability, 64, 68–69
        dissimilarity, 69
        diversity, 69
        efficiency, 64–65, 68
        by importance, 66, **67**, 68–69
        maintainability, 64, 68
        preconditions, 69
        repeatability, 63, 68
        reusability, 63–64, 67–68
        specificity, 64, 68–69
        traceability, 64, 68–69
        usability, 69
    test cases, 61
        factors for development, *70*, 71
Question answering (QA) system, 180; *see also*
        Akinator (game); IBM Watson;
        START; "Twenty Questions" (game)

R

Radermacher, A., 264, 270
Rea, A., 264, 276
ReachGoal method, 295
Recall metrics, 170–171
Recurrent neural network (RNN), 161, *165*,
    165–166, 175
Refined model, 78
Regression testing, 152

Regulative norms, 291
Reliability, 84
Remote pair programming (RPP),
    254
Repeatability, 63, 68
Replacement techniques, legacy systems, 60
Request for quote (RFQ), 128
Request protocol, 297
    interaction diagram, *299*
RESTful API Modeling Language (RAML),
    102, 104
Reusability, 63–64, 67–68
Richardson Maturity Model, 105
Rong, G., 265
Rossi, B., 175
Roy, 254
Roy, N, K. S., 175

S

Salleh, N., 254, 270
Sambasore approach, 204
Scheduling, 93
Scientific Experiments Description Language
    (SEDL), 122
Scripted Collaboration in an Educational Pair
    Programming System (SCEPPSys),
    275
Search-and-selection process, 138–140, *139*
Sequence diagram, 5–6, *6*
Service level agreements (SLA), 128
Sfetsos, P., 262, 270
Shi, W., 204
Simplifieddata.csv, 120, 121
Singh, V., 175
Sison, R., 262
Skip-Gram (SG), 163
Slot Grammar parser, 180
Soft skills, 94–95
Software engineering (SE), 200
Software Evolution Ontology, 203
Software Product Lines (SPL), 108
Software project management
    capable management, 91–93
    functional areas, 93–94
    hard *vs.* soft skills, 94–95
    managing multiple generations, 96–97
    multicultural practice, 95–96, **96**
    poor management factor, 90–91
    project failure attributes, 90–91
    successful team, 97
    training, 95

Software repositories (SR), 201
  automated library, 202
  mining unstructured data, 202–203
  software quality, improvement of, 203
  unstructured data, 202 (*see also* Mining
    unstructured data)
Software testing methods and tools (STMT), 61
Specificity, 64, 68–69
Staffing, 93
START, 181
Sun, X., 204
Support vector machine (SVM), 175
System activity diagram, 221–222, *222*
System Analysis and Mining List
    Recommendation System (SAMLRS)
  analysis of TP, FP, FN and TN, **227**
  ANN-based, *226*
  code executed, determining execution time,
    231
  determining, execution time, *230*
  home page interface, *229*
  performance evaluation, 229
  result, repository link, *230*
  Top-N recommendation algorithm, 227,
    228
  user page interface, *229*
Systematic literature review
  data analysis, 141, **141**
  data extraction, 140–141
  impurity removal, 140
  inclusion and exclusion criteria, 140
  initial search, 139–140
  merge and duplicate removal, 140
  paper selection process, 138–139, *139*
  phases, 137, *137*
  search-and-selection process, 138–140, *139*
Systematic Literature Review (SLR), 253, 278

T

Taji, D., 264
Thematic analysis, 257–258, **258**
Thomas, L., 271
Thomas, S., 204
Thomas, S. W., 202, 204
Thung, F., 204
Tian, Y., 175
Timeit library, 230
Tokenization, 186–187, *187*
Top-N recommendation algorithm, 227, *228*
Traceability, 64, 68–69
Trivial bugs, 163

True negative (TN), 227, *227*
True positive (TP), 227, *227*
Tsompanoudi, D., 274, 275, 277
"Twenty Questions" (game), 180

U

Unified Modeling Language (UML) tool, 203,
    221
Uniformity independent variable, 79
Unstructured data, 200; *see also* Mining
    unstructured data
Urai, T., 274
Usability, 69
Use case diagram, 5, *5*

V

Validity threats, 83–84, 153, 275
  content validity, 72
  data extraction, 72
  population validity
    application age, 72
    technology diversity, 72
Variance inflation factor (VIF), 80
Virtual pair programming (VPP), 254, 274
VisualPPINOT notation, 129

W

Walia, G., 264
Weiss, S. M., 208
Williams, L., 252, 262, 270
Within–subjects factor, 121
Wohlin, C., 275
Wolf, T., 84
Word embedding, 163–164

X

Xia, X., 160
Xinogalos, S., 270

Y

Ying, K., 272

Z

Zacharis, N., 273–275
Zhang, T., 175
Zhong, H., 203